Eye Tracking in Linguistics

RESEARCH METHODS IN LINGUISTICS

Research Methods in Linguistics is a series from Bloomsbury providing introductions to the quantitative and qualitative research methods needed by undergraduate and postgraduate students. The centre of the series is *Research Methods in Linguistics* edited by Lia Litosseliti, which provides a comprehensive overview of all the research methods needed by linguistics students. Each book in the series takes one of the research methods described in the general introduction and expands upon this in a book-length study.

TITLES PUBLISHED IN THE SERIES:

Research Methods for Digital Discourse Analysis, edited by Camilla Vásquez
Research Methods in Linguistic Anthropology, edited by Sabina M. Perrino and Sonya E. Pritzker
Research Methods in Interpreting, Sandra Hale and Jemina Napier
Data Collection Research Methods in Applied Linguistics, Heath Rose, Jim McKinley and Jessica Briggs Baffoe-Djan
Research Methods in Linguistics, edited by Lia Litosseliti
Experimental Research Methods in Sociolinguistics, Katie Drager
Research Methods in Applied Linguistics, edited by Brian Paltridge and Aek Phakiti
Experimental Research Methods in Language Learning, Aek Phakiti

Eye Tracking in Linguistics

SALVATORE ATTARDO
AND LUCY PICKERING

BLOOMSBURY ACADEMIC
LONDON • NEW YORK • OXFORD • NEW DELHI • SYDNEY

BLOOMSBURY ACADEMIC
Bloomsbury Publishing Plc
50 Bedford Square, London, WC1B 3DP, UK
1385 Broadway, New York, NY 10018, USA
29 Earlsfort Terrace, Dublin 2, Ireland

BLOOMSBURY, BLOOMSBURY ACADEMIC, and the Diana logo are trademarks of Bloomsbury Publishing Plc

First published in Great Britain 2023

Cover illustration by Tjaša Krivec

A catalogue record for this book is available from the British Library.

A catalog record for this book is available from the Library of Congress.

ISBN: HB: 978-1-3501-1750-1
PB: 978-1-3501-1751-8
ePDF: 978-1-3501-1753-2
eBook: 978-1-3501-1752-5

Series: Research Methods in Linguistics

Typeset by Deanta Global Publishing Services, Chennai, India

To find out more about our authors and books visit www.bloomsbury.com and sign up for our newsletters.

CONTENTS

LIST OF FIGURES

LIST OF TABLES

PREFACE

This book originates from our experience of teaching both graduate and undergraduate eye-tracking classes at TAMUC over a six-year period. When we started teaching the class, there was no textbook available that would meet the needs of a beginning student, without any background in psychology, and with just a basic understanding of linguistics, such as one could acquire by taking a graduate-level introduction to linguistics.

What we felt was needed was an introduction to the methodology of eye tracking, responsive to the needs of a student without a degree in psychology, starting from the anatomical and neural substrates of vision, an overview of the methodology and the kinds of measurements and experimental setups used in the field and then a robust overview of some of the results achieved in the field of applied linguistics and eye tracking so that the students would know the "lay of the land" so to speak. Finally, we felt it was imperative for those students coming from more humanistic fields or unfamiliar with the process of planning and executing an experimental study to be provided with an introductory treatment of the know-how of doing so, including the statistical treatment of the data.

Luckily we are witnessing a significant shift in the availability of advanced research tools, with both research-grade eye-tracking equipment and advanced statistical software dropping precipitously in price and hence accessibility. While it may still be too expensive for the individual researcher to purchase their own eye-tracking equipment, it may not be unrealistic to think that in thc future we will reach that low entry-level cost. These converging trends have had the effect of popularizing the methodology of eye tracking and at the same time have brought many scholars, some young and others more experienced, in contact with the field, who, however, have no background in experimental research and/or eye tracking. This book is meant to facilitate their introduction to the field. It is not meant to be the last book they will read on the subject but the first one of many. The review questions, list of important words, and the glossary are all tools that we provided with the first-time learner in mind. The glossary serves also as a list of abbreviations/acronyms so that, for example, AAC is found in the glossary under the acronym AAC and the expanded form (Augmentative and Alternative Communication) gets a cross-reference.

Any book is the result of choices and decisions that make the job of producing it manageable. We decided to be concerned primarily with neurotypical populations. While neurodivergent populations, such as autistic and dyslexic persons, are obviously a significant and interesting subject, we decided that these topics were too advanced for an introductory book. We do mention in passing some research on the topic, but we made no attempt at coverage in those areas. Another difficult decision was that of coverage. Some areas (reading, for example) are much more developed than others (social eye tracking, for example). This meant that we could rely on many more syntheses and overviews of the field of reading and eye tracking. This freed us up from having to provide detailed bibliographic documentation of a phenomenon (say, parafoveal preview) that was already covered in detail by several sources, which we made sure to provide for the interested reader. On the contrary, for those areas with less coverage we had to provide more references (in some cases the sole references) on a given subject. This ends up with a rather paradoxical phenomenon whereby the number of sources cited on the most developed areas ends up being less than those cited in the more obscure areas.

We would like to thank Sara Adams, who contributed to the glossary, and Nabiha Al Khatib, who produced several of the images in the text. We would also like to thank all our colleagues at TAMUC who have taught courses on eye tracking with us, and particularly Shulan Lu, Elisa Gironzetti, and Phoenix Carlini, as well as of course the students who took the courses, asked questions, and provided suggestions for improving the manuscript. In particular, John Gibbons provided us with a very helpful set of corrections and suggestions to improve the first draft of the text. Likewise, we owe a great debt of gratitude to the staff of the interlibrary loan office at TAMUC. Needless to say, all remaining errors are our responsibility alone.

We also would like to thank Hannes Ritschel, Daniel Simons, Mario Urbina, and Anke Huckauf, who granted us the right to reproduce images of their work. This is academic collegiality and support at its best.

Last but not least, we would like to thank Andrew Wardell, our commissioning editor, all the anonymous reviewers, and the rest of the staff at Bloomsbury.

The authors and publisher gratefully acknowledge the permission granted to reproduce the copyright material in this book. Every effort has been made to trace copyright holders and to obtain their permission for the use of copyright material. However, if any have been inadvertently overlooked, the publishers will be pleased, if notified of any omissions, to make the necessary arrangement at the first opportunity.

Introduction

If we consider the occurrence of the words "eye tracking" in Google Ngram, as in Figure 0.1, a very clear pattern emerges: the expression is essentially not present in any significant way until the 1960s, which see a first surge of interest, followed by a doubling from the 1980s to roughly 2000 and then an exponential explosion of interest in the following decade.

What explains this remarkable upsurge of interest in the new millennium? After all, the significance of tracking the movement of the eyes to establish what subjects were directing their attention to was already known since the early studies at the end of the nineteenth century and significant results were arrived at in the 1960s and 1970s. Our thesis is that the development of the technology, and in particular the use of computers, has lowered the ingress price of eye tracking on the one hand and the appearance of less intrusive and wearable eye trackers on the other hand have contributed to a widespread adoption of the technology. The cost of a research-grade eye-tracking system has gone down from over \$100,000 to as low as under \$10,000. Eye trackers for gaming are available for under \$100. There even exists free software to do eye tracking using a webcam. Needless to say, the resolution, frequency, and accuracy of these systems vary significantly and only the high-end or research-grade systems are appropriate for research, but just as the availability of light-weight cassette recorders (from 1962, when it was invented by Philips) made it possible to easily record conversations, thus making conversation analysis possible, the increasingly low price of research-grade eye trackers has been responsible for the widespread diffusion of the methodology.

The interest in eye tracking has been reflected in the number of publications on the subject. Titz et al. (2018, p. 1854) showed that the number of articles in Google Scholar with the words "eye tracking" and their variants in the contents of the article increase exponentially beginning around the year 2000. One could possibly read *all* the articles that mentioned eye tracking up until the early 2000s, if one was willing and able to read two or three articles a day, every day for a year, that is. However, by 2019, one would need to read fifty-five articles a day, every day for a year.

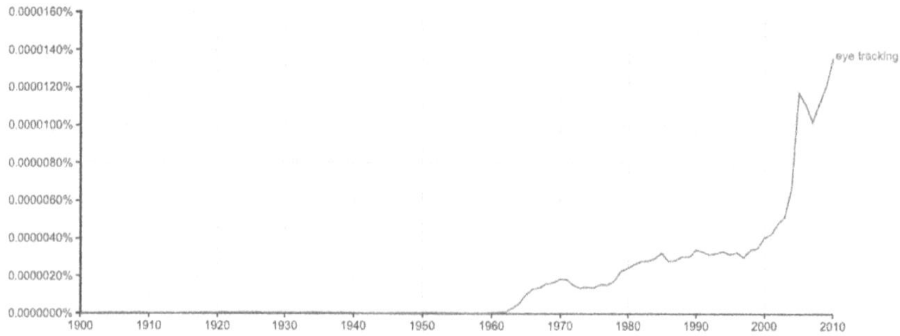

FIGURE 0.1 *Google Ngram for "eye tracking".*

It is because of this interest that this book has come into being, since increasingly both young scholars entering the field of linguistics and older scholars who were trained before eye tracking's "explosion" are interested in using this technology, but they often lack the background necessary to do research in the field. So the purposes of this book are manifold: (1) to give the reader an overview of what research is possible using eye tracking, without obviously attempting to be exhaustive; (2) to provide a step-by-step process of how to set up, execute, and report a study using eye tracking; (3) to provide the necessary background in psychology, physiology, and neurology to understand what the technology is doing/showing.

The book is targeted at scholars who are new to eye tracking or who need an exposure to the theoretical background of the field. However, the book assumes some basic knowledge of linguistics, the kind one would acquire by taking a graduate-level introduction to linguistics. So the glossary does not include terms such as *phoneme* or *morpheme*, which are assumed known. Beyond the glossary, self-study questions are provided for learners who tackle the material independently or to facilitate the work of teachers who adopt it as a textbook in a course on eye tracking.

We should also mention that some attention will be paid to the statistical tools necessary to perform an eye-tracking study. This should not discourage or concern those who come from a humanistic background or whose experiences with statistics have not been entirely positive. Our treatment of the statistical tools for eye tracking requires no mathematical expertise, thanks to sophisticated and freely available software that handles all the calculations, and is focused instead on understanding how to set up the statistical analyses and what they tell us. From a different perspective, one should consider that eye-tracking studies, much like other scientific projects, are complex endeavors and that probably only a research team, comprising several scholars, will have all the necessary expertise to perform the kind of studies that get published. In fact, one of the most striking observations we made while assembling the bibliography for this book is

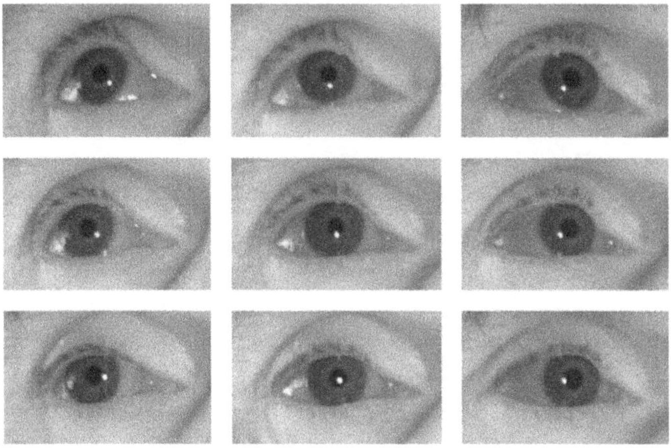

FIGURE 0.2 *The pupil and glint, at different angles of vision.* Source: *Blignaut (2014, p. 2).*

that the vast majority of eye-tracking studies were authored by more than one scholar.

Modern eye tracking is in principle a simple process: a source of light "illuminates" the eye. In many cases, infrared light (which cannot be perceived by the naked eye and is harmless) is used in addition to visible light. A camera records the image of the eye. Then, from the reflection of the light on the eye, we get the **glint.**[1] From the infrared light illumination the pupil becomes clearly visible as a dark circle. Figure 0.2 shows different configurations of the pupil and the glint.

From the position of the center of the pupil and of the glint, it is possible to calculate the line of sight (gaze). The mathematical calculations to get the direction of the gaze from the relative positions of the (center of) the pupil and the glint involve measuring the angles between the camera and the eye, the source of light, and the object being looked at (typically, a computer screen). The eye tracker software performs all these calculations, and once the distance between the eyes and the computer screen is known, it can extrapolate at which point on the screen the user is directing their gaze.

The calibration process, in which the user is directed to look at points whose position is known, has precisely the function of ensuring that the calculations are as accurate as possible, by correcting any distortions, in particular avoiding and correcting known errors of measurement. Figure 0.3 shows two examples of calibration, one not successful (left) and one successful (right): the length of the lines represents the distance between the expected position of the gaze (the five gray circles at the center and the corners) and the observed position.

All modern implementations of eye-tracking technology are based on these principles, but various implementations may differ in the details. For

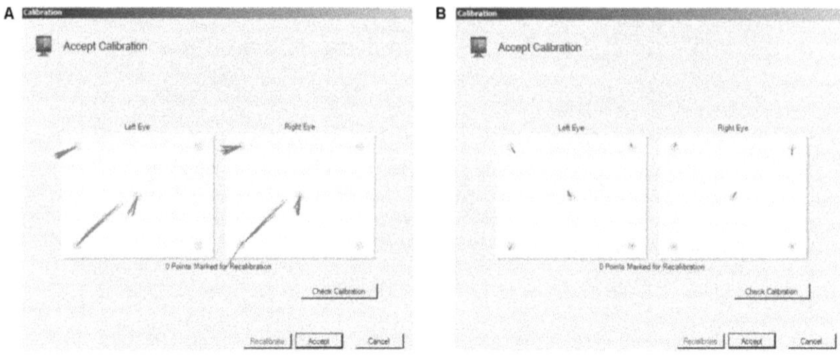

FIGURE 0.3 *Examples of calibration using Tobii Pro Studio.* Source: *Tobii (Dalrymple et al., 2018, p. 2).*

example, more than one source of light may be used, more than one glint may be tracked (usually the first and fourth), the glint *and* the pupil may be tracked, or more than one camera may be used (one to record the position of the eye/pupil and one to record the scene being looked at), the camera(s) may be mounted on glasses, or headsets, and so on. Increasingly, especially for wearable eye trackers, the scene being observed is no longer restricted to a screen, However, the point is that all eye-tracking technology is based on this general principle of calculating the direction of the gaze.

Of course the direction of the gaze is significant because of the other general principle underlying eye tracking, that gaze equals attention. While this is not a necessary correlation, it holds to a surprising degree of detail, despite evidence to the contrary in the difference between overt and covert attention (Chapter 4) as well as in "mind wandering" discussed in Chapter 8.

Eye tracking is a significant addition to the toolbox of researchers who want to understand human (and animal) behavior and thinking because gaze correlates with attention. Eye tracking, as a technology, has emerged in the past century and so it is a relatively new approach and is still evolving, as the trend toward less intrusiveness and smaller footprint shows, but it has already produced a significant body of research and findings.

Part I

The Basics

CHAPTER 1

Historical Development

The history of research on the eye and eye movements goes back to the earliest scientific research of the ancient world. Some notion of the role of eye movements in vision and of the anatomy of the eye was already present in ancient medicine (Babylonian and Greek; Wade, 2010). For example, Galen, a Roman physician (129–210 CE), gave a full description of the muscular mechanisms that move the eyes. Aristotle already mentioned that the eyes move together, and Ptolemy (100–170 CE) wrote a book on optics (which exists in Arabic translation) in which he discussed numerous aspects of vision, including the joint movement of the eyes. The *Book of Optics* by Ibn al-Haytham (aka, Alhazen; 965–1040 CE) contains, among numerous other observations, a description of the optic nerves and the optic chiasm (see Figure 1.1, and Chapter 3 for discussion of the optical chiasm) in the brain.

In our chronological organization we have followed the established four-way periodization (e.g., Cognolato et al., 2018, pp. 2–3): an early phase (up to 1920), a phase of primarily psychological research (until the mid-1970s), a further phase of improvement of the technology (1970–2000), and a current phase, beginning roughly with the new millennium. We would like to stress that no major significance should be accorded to the dates attached to the periods.

The Scientific Pioneers: 1800–1920

Observational Methods

The earliest experimental observations of the eyes' movements date back to the nineteenth century and used self-observations, afterimages, mirrors,

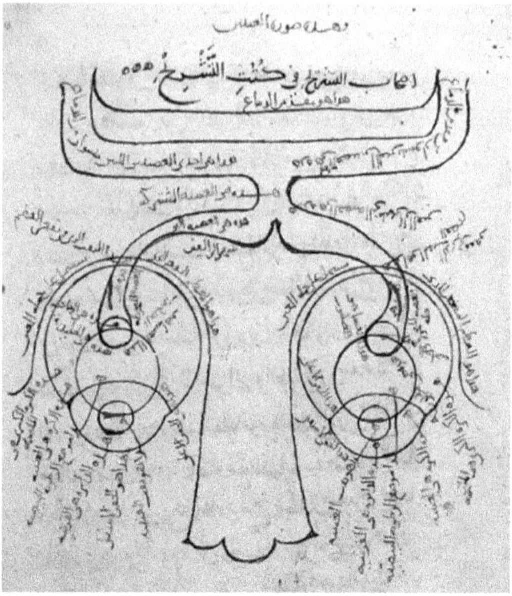

FIGURE 1.1 *Copy of the Kitab al-Manazir (MS Fatih 3212, vol. 1, fol. 81b, Süleimaniye Mosque Library, Istanbul). University of Oklahoma History of Science Collections and BnF Gallica.*

and even a telescope to allow the researchers to observe the movements of the eyes. While these techniques were not precise enough to allow any specific position to be observed, they allowed significant insights. Direct observation, that is, watching the eyes of someone reading, for example, allows the discernment of some ocular movements.

The **Troxler effect** was described as early as 1804. It consists of an observable fading of a visual stimulus in the periphery of the eye if the eye is kept still (either physically immobilized or just by looking in the same spot, say a cross). This is explained by the fact that the eyes are in constant movement and the light receptors on the retina are more sensitive to changes in light than to light per se, so if one keeps the eye unnaturally still, Troxler fading[1] is observed (see Figure 1.2). Indeed, if one moves the eye, even slightly, the faded objects reappear. We will return to this in Chapter 4.

One of the earliest methods to determine eye movements was to use **afterimages.** As is well known, if one looks at a bright object, say a lightbulb, an afterimage remains visible even after the stimulus has been removed (e.g., by closing the eyelids). Using afterimages, it is possible to get a general idea of how much one's eyes have moved relative to an object, since the afterimage is stationary relative to the retina. Needless to say, this method is not very precise. It does have the advantage that it is not intrusive. Use of afterimages goes as far back as 1792 by William Charles Wells (1757–1817). Despite

FIGURE 1.2 *Troxler fading: if you stare at the cross in the center of the image long enough, the fuzzy part of the circle will disappear and possibly the circle itself.*

the coarseness of the method, Wells was able to identify the almost complete blindness that occurs during the **saccade**.[2]

We should stress that these methods (direct observation, afterimages) were not abandoned when other methods appeared. Direct observation was still used in 1878 by Javal, for example. The afterimages method was in continuous use through the 1960s (Wade, 2015). It was used (and named) by Purkinje in 1823. So there is considerable historical overlap in these methodologies.

Acoustic Methods

Ewald Hering (1834–1918) was one of the very first to document saccades (in two 1879 papers) using a rather curious approach: Hueter, in 1878, published an article in which he mentioned that if one applies a tube to the eyelid, a noise can be heard, which he attributed to blood flow in the blood vessels of the eyelid. Hering proved that the sounds were due to muscle contractions, rather than blood flow, as had been hypothesized, by paralyzing the eye using curare (a drug derived from a South American plant that causes temporary paralysis). Since the sounds stopped but the blood flow continued—as the subjects were alive—it followed that the noise was not caused by blood flow. Hering then proceeded to apply the tube to his own eye, which he kept still, but the sounds continued. Using the afterimage technique he was then able to correlate the sounds with saccades. He then applied this technique to reading noting the "jerky movements" of the eye in the process.[3] Hering's work on vision was extensive and included the theory of colors. He is probably best known for his law of equal innervation, which states that since the eyes have the same nerves (innervation) they will move together in a saccade. This was later proven empirically by Yarbus. Further discussion will be found in Chapter 4.

Louis Émile Javal (1839–1907) was one of the leading ophthalmologists in France. In 1877, he was the founding director of the ophthalmology laboratory at the Ecole Pratique des Hautes Etudes. Among his collaborators was Lamare.[4] Lamare established that readers did not follow the line of text smoothly and continuously. He also described the jerky movement of the eyes, and he and Javal used the term "saccade," which has become a technical term. According to Lamare, in 1879, he developed a technique to count saccades by amplifying the sound produced by each saccade: a small tube was attached to the eye (either the eyelid or the external membrane of the eye) and thus someone listening through a tool similar to a stethoscope could easily count the saccades. However, Lamare failed to publish his results for ten years. Thus the discovery of saccades is generally attributed to Javal (Wade et al., 2003; Pluzyczka, 2018). However, Javal never claims the discovery for himself and quotes Lamare's work at length.[5]

Summing up, in the 1878–9 period, Hueter in Germany, Hering in Prague, and Lamare in Paris were using roughly the same technique to investigate saccades in reading. Truly an example of an idea whose time had come! However, while the auditory technique allowed one to count saccades and to distinguish the shorter saccades during the reading of a line from the longer ones when the reader goes to the next line, it did not provide a visual documentation of the phenomenon.

Visual Methods

The earliest visual methods involved photosensitive plates (like those used in photography). Stratton (1902, 1906), using this technique, documented the irregular movements of the eye over images, including visual illusions. His work anticipated that of Buswell and Yarbus (see later in the text).[6] Several other researchers applied this method (see Vernon, 1928), which however had the inherent limitation that it can only document the movements of the eyes but not the duration of those movements.

Dodge (1907) developed the first videographic apparatus for tracking eye movements. He came up with the idea of illuminating the cornea with a "pencil of light" and recording its reflection with a video camera (then called a kinetoscope), which had just been invented around 1890 and commercialized a few years later. The device relies significantly on a "head-holder" (1907, p. 92) visible in Figure 1.3.

Edmund Burke Delabarre (1863–1945) took some plaster of Paris casts of an artificial eyeball and trimmed them so that they would fit on an eye. He then anesthetized the eye using a 3 percent solution of cocaine[7] and attached the cast to the eye. Delabarre experimented with a small mirror,

FIGURE 1.3 *The Dodge Photochronograph.* Source: *Dodge (1907).*

which reflected a beam of light and thus produced a record of the movements of the eye on a light-sensitive plate. However, he found that the result was not satisfactory, as the movement of the light was too fast for the plate to capture it adequately (Delabarre, 1898).

Edmund Burke Huey (1870–1913) wrote his PhD dissertation on reading (1899). In Huey (1898), he described the apparatus (see Figure 1.4), similar to that of Delabarre, used to confirm that eyes do not move smoothly but in saccades. He later published a book (Huey, 1908) on reading, which had a major impact on the field of vision. Huey is of particular interest to the history of eye tracking, because he originated the misattribution of the discovery of saccades. Huey had visited Javal's laboratory, and he attributed the discovery of saccades to Javal. Common opinion on the matter adopted Huey's view, until Wade's work (Wade et al., 2003; Wade & Tatler, 2005, 2009; Wade, 2010) set the record straight. An excellent account of Huey's life and work is Reed and Meyer (2007).

Charles Hubbard Judd (1873–1946) developed a method consisting of placing a Chinese white[8] speck on the cornea as a registering mark and then using a video camera (kinetoscope) to film the movements of the eyes. He was thus able to document that both eyes move together (binocular convergence) unless one deliberately moves them apart. In order to eliminate head movement, the subject clenched their teeth around a bar and later the object looked at was attached to a bamboo reed held in the mouth of the subject (Judd et al., 1905; Judd, 1907). Figures 1.5 and 1.6 show the mechanisms (the subject in Figure 1.6 is Judd himself). The wire "spectacles with steel beads," visible around the eyes, are fastened to the head and provide another set of reference points. In the figures the lower white dot is

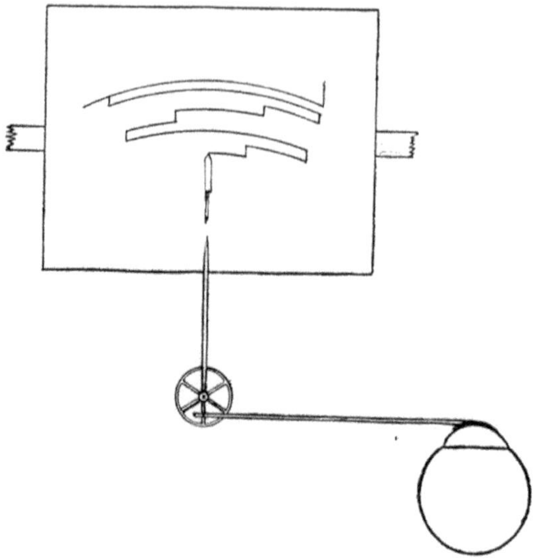

FIGURE 1.4 *The apparatus used by Huey (1898, p. 584). The circle in the bottom right of the image is the eye, with a plaster cup on it, to which is affixed a small rod, which is connected to a recording device. Delabarre's apparatus was similar.* Source: *Huey (1898, p. 584).*

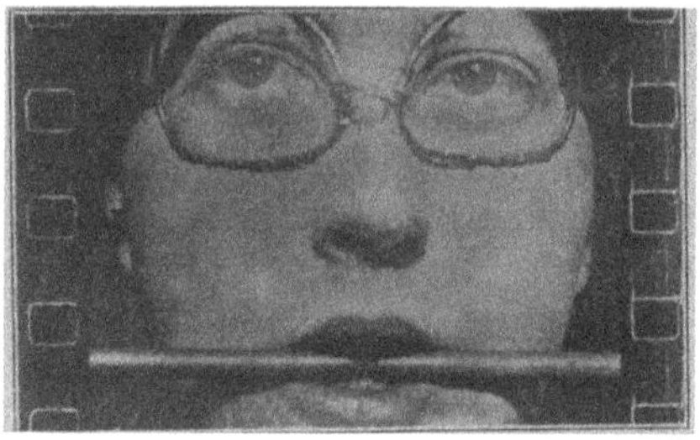

FIGURE 1.5 *Judd's original apparatus.* Source: *Judd, McAllister and Steele (1905).*

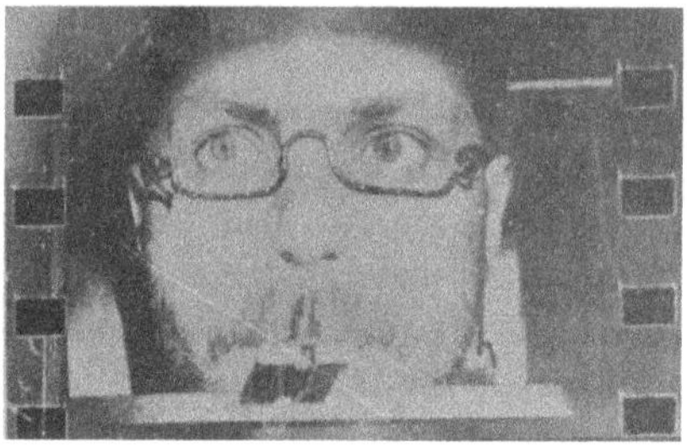

FIGURE 1.6 *Judd's revised apparatus.* Source: *Judd, McAllister and Steele (1905).*

the Chinese white speck, whereas the upper large white spot is the reflection of the light.

In the first quarter of the twentieth century several researchers explored the idea of measuring eye movements by attaching some kind of marker to the cornea of the eye itself and then recording the movement of the marker. The simplest method was to immobilize the head of the subject, anesthetize their eye, and then proceed to rest a small level on the eye. The movement of the cornea was then recorded on a paper tape. A similar technique used a rubber

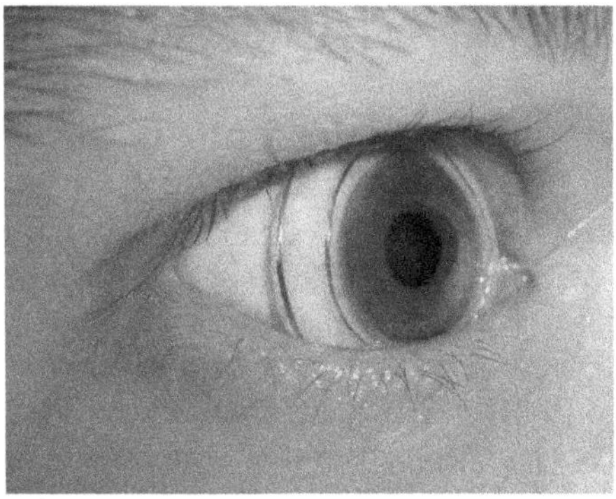

FIGURE 1.7 *Scleral lens with coil.* Source: *CHRONOS VISION GmbH, reproduced with permission.*

balloon, also in contact with an anesthetized eye. The balloon controls the output of an acetylene flame, which produces a smoke mark on a strip of paper (see Vernon, 1928, p. 125, for a full description and schematic representation of the apparatus, attributed to Schackwitz, 1913). Finally, a small cup, similar to a contact lens was employed, to which a level or thread was affixed. The movement of the latter was then recorded. Needless to say, this method too required anesthetizing the eye. Initially these cups were made of plaster of Paris (following Delabarre's method) but later ivory, aluminum, and regular contact lenses were also used. In some cases the mirrors were attached directly to the eye (the sclera, to be precise). Yarbus's caps were a significant upgrade because they were made of rubber. The fact that in most of these techniques the subjects' eye needed to be anesthetized is a stark reminder of the degree of intrusiveness that these methods required.

Along the same lines, various researchers used a small mirror to reflect light, which could then be recorded. The mirrors were affixed to the eye using a variety of methods, including contact lenses and suction caps. The contact lenses would seem the logical method to attach a tiny device to the eyes, but they have the problem that they only fit loosely and they move during regular vision, making them less effective for precise measurements. Suction devices were perfected by Yarbus (see later in the text). Later, scleral contact lenses with tiny wire coils embedded in them were produced (Robinson, 1963). **Scleral lenses** are bigger than regular contact lenses, as they rest on the sclera and not the cornea. Figure 1.7 shows a scleral lens with coil. The model is wireless, thus affording extra convenience.

The **scleral coils** produce a magnetic field that can be detected and used to track the movement of the eye in three dimensions. This method is still in use currently, especially in situations in which optical eye tracking is not practical.

Applied Behaviorist Research: 1930–70

Image Perception

Guy Thomas Buswell (1891–1994) was an educational psychologist who performed a series of pioneering studies using newly developed techniques in eye tracking. His technique is credited as being the first nonintrusive eye-tracking technique. Buswell's primary area of interest was reading; for example, he showed that when reading aloud we look ahead of the words we are actually saying. He then turned the techniques developed for reading to the perception of visual images. His 1935 book *How People Look at Pictures* is considered a classic in the field.

Buswell noted that, when looking at an image, viewers orient their attention toward "centers of interest" (p. 18) and the distribution of fixations is far from uniform or random. He further found a significant difference between early, more predictable fixations, and later, less predictable ones (pp. 27–45). Buswell (1935, p. 142) identified two "general patterns of perception" when looking at pictures of paintings: (1) an early "general survey" consisting of a series of short fixations over the "main portions" of the picture; and (2) a later series of longer fixations "concentrated over a small area of the picture, evidencing detailed examination of those sections" (p. 142). Buswell characterized the two patterns as "survey" and "detailed study," respectively. This distinction has been adopted by Yarbus and his collaborators (Karpov et al., 1968) and many others, for example, Antes (1974), Babcock et al. (2002), Liechty et al. (2003), and Pieters and Wedel (2004). Finally, Buswell noted that the instructions provided to the participants "have a marked influence upon the character of perception" (p. 144).

Buswell's apparatus (Figure 1.8) was especially developed for the purpose of the study of image perception. It exploits the reflective capacity of the cornea, on which a tiny beam of light is projected and then recorded on film. The participants are seated at a table on which the apparatus is placed and their heads are held in place by a head mount. The size of the apparatus is that of a large table. Since the light source and the recording equipment are to the side of the participant, the setup allowed for the display of various images of varying size at various distances from the seated participant.

Yarbus (1967) is both a synthesis of previous research as of 1967 and the first systematic studies involving complex scenes. With Yarbus, eye-tracking methodology reached a certain maturity: the central ideas are well established and can be considered solid results and taken as a starting point for further more detailed research.

Yarbus presented in significant detail his method of using small suction devices (caps, 1967, pp. 29–40) that are affixed to the eye and because of the small pressure differential remain firmly attached to it. However, the eyelids

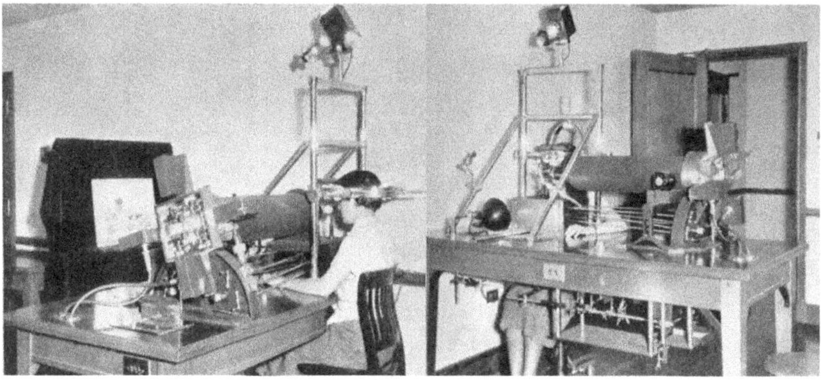

FIGURE 1.8 *Buswell's 1935 eye-tracking apparatus, front and rear views.* Source: *Buswell (1935).*

have to be restrained (using adhesive strips of plaster of Paris) to prevent blinking, which would dislodge the caps, so the devices can work only for a few minutes at a time, which is indicative of the level of discomfort the subjects must endure. Moreover, the caps, by Yarbus's own admission, produce "irritation" of the conjunctiva (a part of the eye, see Chapter 2). As we saw earlier, a small mirror is affixed to the cap, and by shining a light beam on the mirror, the movements of the eyes were recorded on photosensitive paper.

Yarbus started out by exploring and documenting the Troxler effect and established that constant eye movement is essential for vision. He came to the conclusion that the impression of a steady "stationary" fixation on an object is an illusion: the eyes move in (1) microsaccades, matching in both eyes; (2) drifts, that is, small random slow movements, which, however, remain within the fovea (p. 108); and (3) tremor (now called ocular microtremor), a high-frequency twitching (trembling) of the eye. These micromovements are completely normal. In fact a diminution of the frequency of microtremor has been used to diagnose multiple sclerosis (Bolger et al., 2000). Conversely, saccades are very fast and the eyes move synchronously. The biggest distance of a saccade is 20 degrees. The duration of a saccade is between 1 and 7 ms (p. 146).

Yarbus is now remembered primarily for his work on complex scenes, which was truly pioneering. He used a wide variety of images, but his study of the painting by I. E. Repin, *An Unexpected Visitor*, is given more space and has gathered the most interest. The painting is reproduced in Figure 1.9. His first conclusion was that the points that one will fixate in an image will vary depending on the kind of information one is looking for. As Buswell (1935) had already noted, the instructions made a significant difference in the processing of the image. So, when Yarbus showed a painting to his

FIGURE 1.9 *I. E. Repin's* An Unexpected Visitor, 1884–1888.

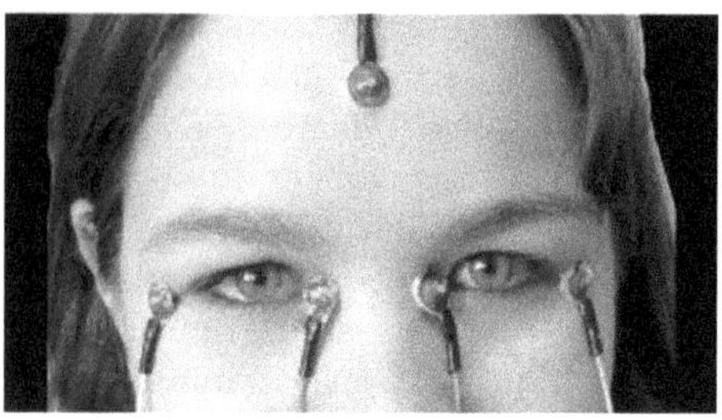

FIGURE 1.10 *An example of electrooculography.* Source: *Metrovision, reproduced with permission.*

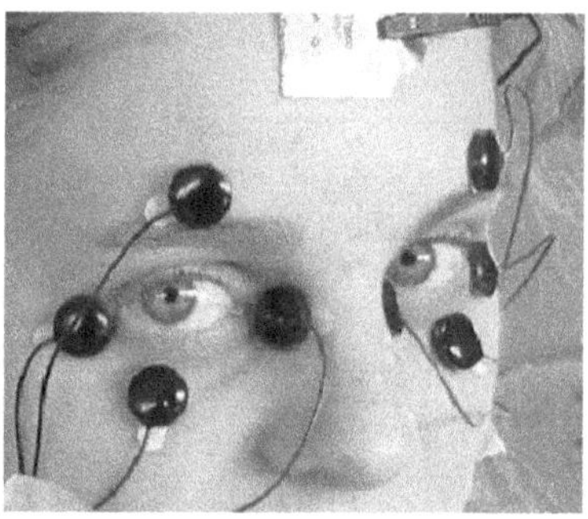

FIGURE 1.11 *Example of electrooculography with four electrodes per eye, to better record both vertical and horizontal movements.* Source: *Metrovision, reproduced with permission.*

subjects and gave them directions to assess the "material circumstances of the family" the observer paid more attention to the furniture and clothing, whereas when the instructions were to guess what they were doing before the arrival of the visitor, the observer looked at the object on the table, the music, and the hands of the characters (pp. 171–96).

Yarbus summed up his research in the viewing of complex images as follows: "the distribution of points of fixation on the object changes depending on the purpose of the observer" (p. 211).

Other techniques developed during this period include **electrooculography**, that is, the use of electrodes placed around or across the eyes, to detect the differences in electric potential between the points (see Figures 1.10 and 1.11). A related approach, also developed in this period, consists of measuring the electrical differential between the cornea and the retina (see Young & Sheena, 1975). Given our focus on visual methods of eye tracking, we will not consider these approaches in any detail.

1970–2000: Improvement on the Technology

In 1974, Lambert et al. introduced an eye-tracking system (then called an "oculometer") that had the great advantage of not requiring the user to be strapped in the machine and thus allowed far greater flexibility in data collection. However, the system was literally the size of two rooms (the subject sat in an anteroom, with the camera hidden in the base of a display screen; the operator and the computer to record the data were in an adjacent room). Mainframe computers were the size of large closets at the time.

The biggest improvement to the optical methods of tracking the pupil was to use infrared light: infrared light is absorbed by the eye, making the pupil darker than the iris, which allows for greater visibility. Corneal reflection was also improved in the late 1950s by an apparatus that allowed the superimposition of the spot reflected by the cornea on the stimulus seen by the subject (Young & Sheena, 1975, p. 403). The size of eye-tracking equipment decreased significantly; while the earliest machines were self-built and sometimes the size of a room, eye trackers began to become smaller and easier to handle. Indeed, during this period the first head-mounted cameras appeared. Technically, the first such apparatus was presented in Hartridge and Thompson (1948), but it uses a mouth bar. Shackel (1960) is an example of an early head-mounted camera/eye tracker, which uses a helmet to which the camera is secured. Mackworth and Thomas (1962) used a head-mounted camera, with a little periscope to reach the eye. Other examples are shown in Young and Sheena (1975, pp. 404–5). The obvious problem is that "the interference with head movements associated with the weight of the instrument may be considerable" (Young & Sheena, 1975, p. 404). During the 1970s further applications of eye trackers to helmets, with the use of transmitters, to reduce the weight problem, were developed to be used in flight and military settings.

Land and Lee (1994) developed a head-mounted eye tracker that also recorded the scene being looked at, thus allowing the superposition of the field of vision and of the eye movement. They applied it specifically to driving a car, and they showed that drivers look at the tangent point (Figure 1.12) when handling a turn in the road.

However, with few exceptions, all these eye trackers required the limitation of head movements. This can be achieved by using two possible methods: on the one hand, one can restrict the head movements by placing the subject in a chin rest, which prevents all but the smallest movements of the head (Figure 1.13). On the other hand, one can ensure that the head and the equipment move in the same way using a bite bar (see Figure 1.5, for an example).

Head stabilization is still used for applications that require extreme precision, such as the study of how diacritics affect reading in Arabic (Hermena et al., 2015), where a nonstabilized eye tracker would not be precise enough, or the study of rapid phenomena that need to be captured at a higher sampling rate, such as tremors (see Chapter 4).

While they do solve the problem of head movements, these devices are hardly comfortable and prevent the collection of naturalistic data, let alone

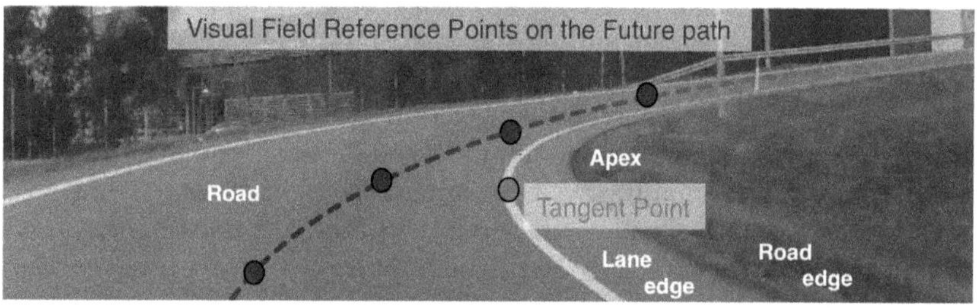

FIGURE 1.12 *Illustration of the tangent point.* Source: *Itkonen et al. (2015).*

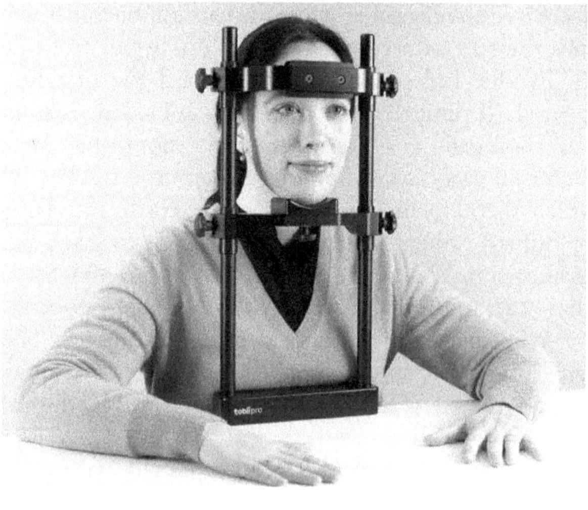

FIGURE 1.13 *An experimental participant in a chin rest.* Source: *Tobii, reproduced with permission.*

FIGURE 1.14 *Eye-tracking glasses: Tobii-Pro 3 glasses.* Source: *Tobii, reproduced with permission.*

of social situations (e.g., conversations). It is hard to talk while biting down on a metal bar. This sort of problem was addressed with the next generation of eye trackers, when wearable technology was introduced.

Wearable Eye Tracking: 2000–Present

Summing up, in the early 2000s, Bates et al. (2005) and Cognolato et al. (2018) identified several distinct methods of eye tracking:

1. Video **oculography**: Light reflection of the pupil and cornea or just of the pupil. Other options include tracking the **Purkinje images** (corneal reflections)[9] and the limbus (i.e., the iris-sclera boundary; see Chapter 2).
2. Electrooculography: The measurement of the voltage (electric potential)[10] across electrodes placed on the skin around the eye; see Figures 1.10 and 1.11.
3. Electromagnetic: Scleral coil in the eye/corneal lens; see Figure 1.7.
4. Contact lenses: Contact lens in the eye (tracking the contact lens).

There is a general consensus that "by far the most popular gaze-tracking methodology was video oculography using pupil and corneal reflection" (Bates et al., 2005, p. 19; Cognolato et al., 2018, p. 3). Often, the accuracy of video oculography is further improved using infrared light, which is particularly useful to resolve eyelid covering problems (e.g., blinking).

Cognolato et al. (2018) reviewed head-mounted (wearable) eye-tracking devices, in the context of providing support to persons with various impairments. They reported that they were able to find six head-mounted or wearable eye trackers that are commercially available. Two of them are glasses.

Cognolato et al. (2018) described the fourth "era of eye tracking" as "supported by the technology evolution in telecommunications, electronic miniaturization, and computing power" (p. 3). We see this trend in full force in Figure 1.14, which shows eye-tracking glasses essentially indistinguishable from regular glasses. The front-view camera is built in the space between the lenses. The infrared illuminators for the eyes and two cameras are built in the lenses (they are the small dots you can see under the eyes of the model). The glasses collect the images of the eyes using two cameras for each eye and an image of the field of vision. They also record ambient sound with a microphone located above the camera in the frame. The glasses weigh 76 grams (2.7 ounces), a far cry from the head-mounted cameras introduced in the 1970s.

The data are collected onto an SD card housed in a battery unit worn in the pocket, which gives over 100 minutes of recording time, per charge. This allows truly untethered collection of data, outside of the laboratory. The most significant advantage of eye-tracking glasses is that they allow the use of eye-tracking technology in naturalistic settings. The potential for marketing and other practical applications is obvious.

The current tendency for wearable eye trackers is toward glasses, as evidenced by the consumer interest in "smart glasses" (without eye tracking). The way smart glasses work is that they incorporate a display on which a built-in computer can display information relevant to the situation. A front-facing camera records the field of vision and can be used to search for relevant information (e.g., faces to be recognized, or street signs, etc.). However, privacy concerns led to Google glasses sales to the general public ultimately being discontinued, in 2015. They have since produced another version and there are many other "smart glasses," but they are mostly targeted at industrial and gaming applications.

The availability of glasses that record the scene facing the user makes the step of blending eye-tracking glasses and smart glasses fairly predictable; it could be used for applications such as human-computer interface (**HCI**), for example, for users of augmentative and alternative communication (**AAC**) devices or for situations in which one's hands are otherwise occupied. Eye tracking is already used in gaming, so that also looks like a logical direction for the field to expand. We discuss these topics in Chapter 9.

Finally, looking at the future ahead, a recent development is the application of deep learning to eye tracking. The basic idea is that by learning from a very large set of images a neural network can be trained to identify the position of the pupil in images and hence the gaze (e.g., Krafka et al., 2016; Yiu et al., 2019). This could lead to eye-tracking capacity in a simple smartphone.

Conclusion

We see two related trends in the development of eye-tracking technology: a trend toward greater ease of use and less intrusiveness, on the one hand, and a tendency to incorporate new technologies as soon as they become available, on the other hand. For example, within ten years of its becoming commercially available, Dodge (1907–1908) employed video recording to record pupil location. Note that still photography fails to capture the all-important temporal dimension of eye tracking. It takes video recording to implement effective eye tracking. However, this is still an analog medium. It will take the digital computer for eye tracking to come to maturity in the 1960s–1970s.[11] The usefulness of the computer does not lie only in its capacity to record data but more importantly in its capacity to process it. Modern eye-tracking equipment provides the researcher with predigested data in which fixations and saccades have already been segmented (for details of the methodology, see Chapter 4). Prior to the advent of computers, this process had to be done by hand. Finally, technologies such as Bluetooth and Wi-Fi allow data streams to travel wirelessly from headsets or glasses to the computers that do the processing of the data.

The trend toward less intrusiveness is fairly easy to discern, and researchers comment on it occasionally. The less intrusive methods (unaided direct observation of the eye) are not precise enough, but precision meant, until the fourth period, intrusive methodologies, with apparatus directly applied to the eye, or in the form of head restraints, to prevent head movement. With the advent of remote imaging of the pupil (with or without infrared light), the physical intrusiveness was all but resolved, but, nonetheless, the size of the apparatuses used and the need to calibrate the equipment still impeded any naturalistic data collection. It is only with the recent development of small, portable, and wearable eye-tracking devices that it has become possible to collect naturalistic data, for example, someone walking down the aisle of a supermarket, or even conversational data (see Chapter 7) with more than one participant.

Self-Study Questions

The following questions pertain to the contents of this chapter. You should be able to answer them without looking up the answer in the book.

1. Rank the various approaches to eye tracking in terms of intrusiveness.
2. Explain why the application of eye tracking to driving and flying planes only started in the 1970s.

3. Explain the purpose of chin rests and head restraints.
4. Explain the difference between oculography and electrooculography.
5. What are the advantages of glass-mounted eye-tracking equipment?
6. What is the advantage of using infrared light in addition to visible light in eye tracking?

Words to Know

The following is a list of words that have been introduced or discussed prominently in this chapter and that you should be able to define. You may test yourself on them, for example, by making flash cards.

Troxler effect	afterimages	saccades	HCI	AAC
oculography	scleral coil	scleral lens	Purkinje images	
electrooculography				

Further Reading and Sources

The discussion of the history of eye-tracking technology is based on Vernon (1928), Young and Sheena (1975), Bates et al. (2005), Morimoto and Mimica (2005), Wade and Taler (2005), Cognolato et al. (2018), and Pluzyczka (2018).

CHAPTER 2

Anatomy and Physiology of the Eye

This chapter deals with the anatomy of the human visual system, that is, how the various organs that compose it are made up, and with its physiology, that is, how it functions. Technically, the visual system is a part of the central nervous system, but in this chapter we will be concerned mostly with the anatomy and physiology of the eye. The next chapter addresses vision and the role of the brain in the process.

Schematically, the human vision system can be represented as in Figure 2.1 below. The image is seen from above. A version from the side, which gives a better understanding of where the organs are located, is found in Figure 2.2. The vision process can thus be considered as the transfer of photoreceptors, information to the back of the brain for processing by the visual cortex.

Anatomy

The eye can be visualized as a sphere with a bulbous extension. It can be divided into a front and a back part, divided by the iris and the lens. Functionally, the eye works like a camera. An object reflects light; the light enters the eye and is focused by a lens on the back wall of the eye. Figure 2.3 shows schematically the basic functioning of a camera. Note that the image is reversed on the photo sensor or film.

Sclera

The **sclera** is the white, opaque external part of the eye that is visible (the "white of the eye"). It makes up 85 percent of the outer "tunic" of the eye; the other 15 percent is the cornea. It is made of collagen fibers, much like the skin,

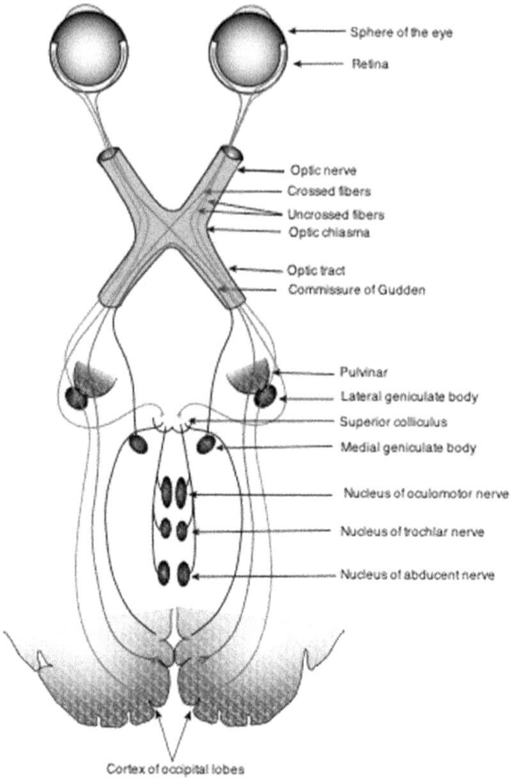

FIGURE 2.1 *The human vision system.* Source: *KDS4444, Wikimedia Commons.*

and its main function is to provide and maintain the shape of the eye. The muscles that move the eyes are attached to the sclera. The sclera is continuous to the cornea. The boundary between cornea and sclera is the **limbus.**

Cornea

The cornea is the transparent part of the eye. It bulges slightly out. It consists of six layers. The outmost layer consists of a very thin layer of cells, the epithelium, that are kept moist by tear fluid. Blinking ensures the distribution of the tear fluid on the cornea. The tear fluid is produced by the tear (lacrimal) glands, which are located above the eyes. The other layers of the cornea, also transparent, consist mostly of collagen. The cornea's main physiological function is to focus an inverted image on the retina. The cornea is heavily innervated, and thus very sensitive to pain, which is why a scratch to the cornea is extremely painful. If touched it triggers an involuntary reflex to close the eyelid. Lasik surgery can reshape the cornea

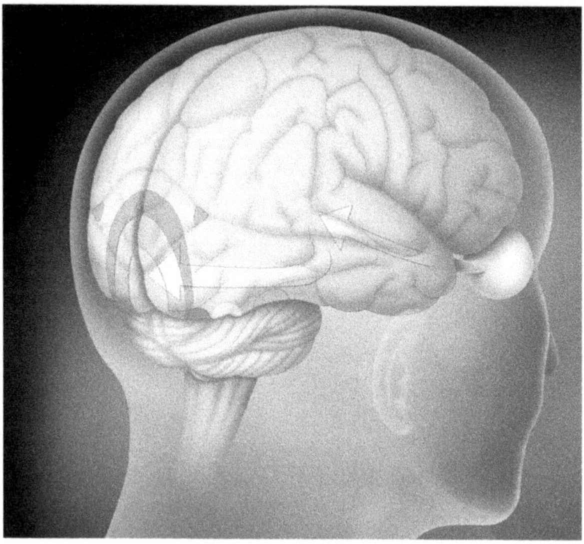

FIGURE 2.2 *The human vision system (side view); information travels from the eyes to the visual cortex.* Source: *Science Photo, reproduced with permission.*

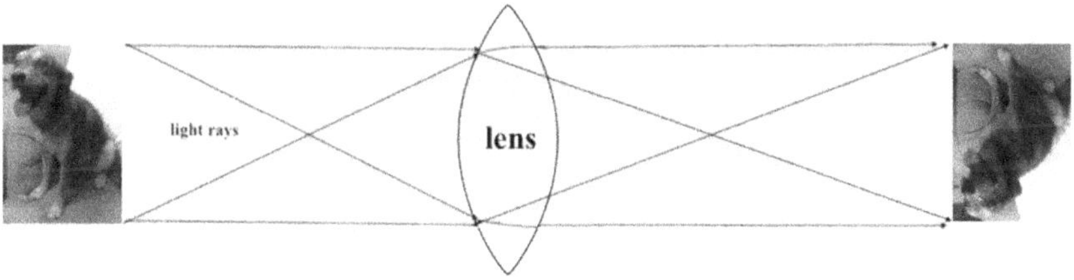

FIGURE 2.3 *The basic structure of a camera. The image is inverted on the back of the camera.* Source: *Science Photo, reproduced with permission.*

using a laser when it no longer focuses the image properly, due to vision problems, such as myopia (nearsightedness), or aging.

Conjunctiva

The conjunctiva is a transparent mucous membrane that covers the eye ball and the inside of the eye lids but not the cornea. You may be familiar with the term because of conjunctivitis ("pink eye"), that is, an infection of the conjunctiva which causes reddening and suppuration. The function of the conjunctiva is to protect, moisten, and lubricate the sclera and to nourish the cornea. You can think of it as a protective layer keeping dust

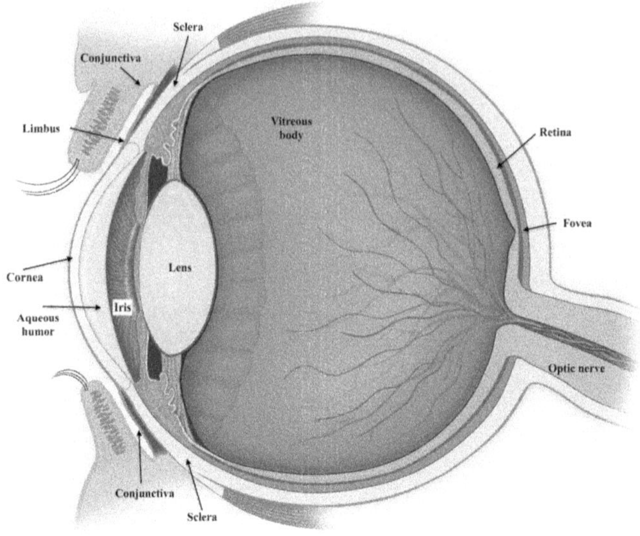

FIGURE 2.4 *The anatomy of the eye.* Source: *Science Photo, reproduced with permission. Modified by the authors.*

and microorganisms out of the eye. Figure 2.4 shows the conjunctiva above (palpebral) and below (bulbar) the eye, respectively, in relation to the cornea and the sclera (in light gray).

Iris

The iris is a thin membrane separating the front and the back of the eye. The pupil is the central opening of the iris. The iris is the colored part of the eyes. When we say that someone has green or brown eyes, we mean that their iris is that color. Eye color is determined by the amount of melanin (the brown pigment in dark skin) in the iris and by light scattering, in an effect similar to the blue "color" of the sky: much like the sky is not blue, but it appears to be so due to light scattering, there are no blue or green pigments in the eyes but they appear to be that color due to light scattering. For example, in blue irises, there is little or no melanin in the iris and so the light passes through it, but its relative density causes scattering and hence the blue color.

Pupil

As we saw, the pupil is the opening of the iris. The size of the opening can range from 2 to 8 mm under normal conditions. The dilation or closing of the pupil is determined by two muscles in the iris. The function of the pupil is to determine the amount of light entering the back of the eye. Increased illumination causes the pupil to constrict. Besides the amount of light, pupils may be dilated or constricted by emotions and constricted by sleep. Generally speaking, both pupils in one individual have roughly the same size; however, about 20 percent of the population demonstrate **anisocoria,** that is, pupils of different size. The difference is usually less than 1 mm. Medical conditions and the use of drugs may also cause anisocoria. This is significant when doing pupillometric studies, that is, studies that measure the dilation of the pupil. The influence of luminosity on pupil dilation and constriction is the reason that care should be taken to have consistent neutral illumination when performing eye-tracking studies. When focusing on a nearby object the pupil also constricts.

The process of the pupil's contraction and dilation is called the **pupillary light reflex.** The ganglion cells in the retina are sensitive to the quantity of light. They send a signal to the oculomotor nerve, which, as the name indicates, has the function to move the eyes and that causes the muscles in the iris to open or close the iris and thus make the pupil bigger or smaller. The dilation or constriction of the pupil is not just governed by intensity of light. Cognitive factors are also relevant. For example, arousal or the fight-or-flight response has been shown to cause dilation of the pupil. Some drugs may also cause dilation or restriction of the pupil. **Pupillary hippus** is a spasmodic contraction and dilation of the pupil, which is not related to illumination or eye movement and is usually normal. The opening and closing of the pupil is a relatively slow process, which takes place at a visible rate (e.g., pupillary hippus takes one or two seconds). Pupil dilation due to mental effort or arousal takes seconds (Klingner et al., 2008). After light is perceived, it takes a certain amount of time before the pupil can start contracting. This is called **latency**. Latency is shorter as the intensity of the light increases. The shortest latency measured is 180–220 ms (Hall & Chilcott, 2018).

In conclusion, the interdependence of light conditions and pupil dilation makes it crucial that eye-tracking studies and particularly studies involving the measurement of the pupil's diameter (pupillometry) be performed under uniform light conditions.

Lens

The lens is, as the name indicates, the transparent element of the eye that focuses the light rays onto the retina. It sits between the aqueous humor and the vitreous body (see Figure 2.4), immediately behind the iris and the pupil. It is held in place by two sets of suspensory ligaments, attached at its widest point (known as the equator of the lens). The ligaments themselves are attached to two muscles (the ciliary muscles) Normally, the eye is focused on remote objects. If you focus the eye on an object that is very close a process called accommodation takes place to facilitate vision. The ciliary muscles act on the lens and deform it slightly, thus changing its focus and accommodating for the vision of the near object. Other animals, such as horses, cannot do this and as a result have to tilt their heads so as to change the focus.

As we saw the lens is transparent. However, diseases, such as diabetes, or just aging may cause the lens to lose its transparency and become opaque. This is called a cataract. Surgical procedures can be performed to remove a cataract and replace it with a clear lens.

Purkinje Reflections

As we saw in the introduction, the incoming rays of light are reflected by some of the structures of the eye, thus creating reflections, called Purkinje images. We are now in a position to describe the process in slightly more detail: essentially, the cornea and the lens reflect the incoming light (both visible and infrared). Since there are two surfaces (front and rear) for each structure, we get four images. Purkinje images are usually labeled: P1(front of the cornea), P2 (back of the cornea), P3 (front of the lens), and P4 (back of the lens). Figure 2.5 gives a schematic representation of the process.

P1 and P4 are usually used for eye-tracking purposes: P1 is the clearest and strongest; P4 is only 1 percent as bright. P2 is very close to P1 and very difficult to see; P3 is larger and fuzzier, as it forms in the vitreous humor (see The Aqueous and Vitreous Humor section).

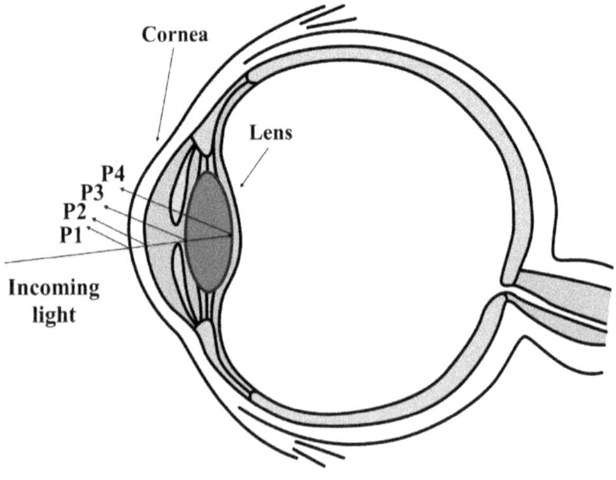

FIGURE 2.5 *The Purkinje reflections.*

However faint, P4 forms in the lens and because of the concave shape of the lens, it is inverted (unlike P1–P3). If the eye moves in a straight line (translational motion), P1 and P4 move together. If the eye rotates, they do not. As a result, their relative motion can be used to calculate the angle of the motion.

The Aqueous and Vitreous Humor

The aqueous humor is located in the front part of the eye (anterior chamber), between the cornea and the iris. As the name indicates it is essentially water and thus transparent. Its function is to convey oxygen and nutrients to the lens and the cornea, to carry away any waste or inflammation from the anterior part of the eye, and to preserve the shape of the eye. The eye regularly produces a constant amount of aqueous humor and some of it drains off. If there is an imbalance of production/drainage the pressure in the eye can increase and this can cause glaucoma, which may damage the optic nerve and cause blindness.

The vitreous humor is a transparent gel, composed 99 percent of water and also collagen fibers, sugar, salt, and ascorbic acid. Its function is to carry nutrients to the lens and retina. On average the human eye has about 4 cubic centimeters of it. The vitreous humor is not replenished, so losses can lead to vision problems such as retina detachment or macular holes.

Retina

The retina lines roughly the back two-thirds of the eye. Functionally, it corresponds to the light sensors or the film in a camera: it is where the focused image produced by the lens is projected. The retina is responsible for light/dark and for color vision. There are three types of photosensitive cells (i.e., capable of perceiving light): the rods, the cones, and the ganglion cells. They are examined in more detail in Photoreceptor Cells section.

Within the retina we distinguish two areas: the macula and within the macula, the fovea. The macula is a darker pigmented area of the retina of about 5 mm (or about one-fourth of an inch). It is responsible for color vision. The macula and the fovea are visualized in Figure 2.6. The dark filaments are blood vessels.

Fovea Centralis

The *fovea centralis* (Latin for "central depression"), or just **fovea**, is, as the name indicates, a depression in the retina in the macular region. It has the highest acuity of vision and is the area where the image is focused, when the eye is functioning normally. In myopia (nearsightedness) or presbyopia

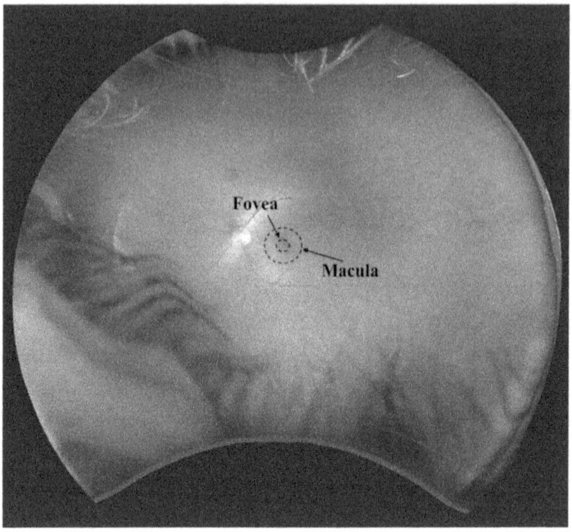

FIGURE 2.6 *The macula and the fovea of the left eye of one of the authors.*

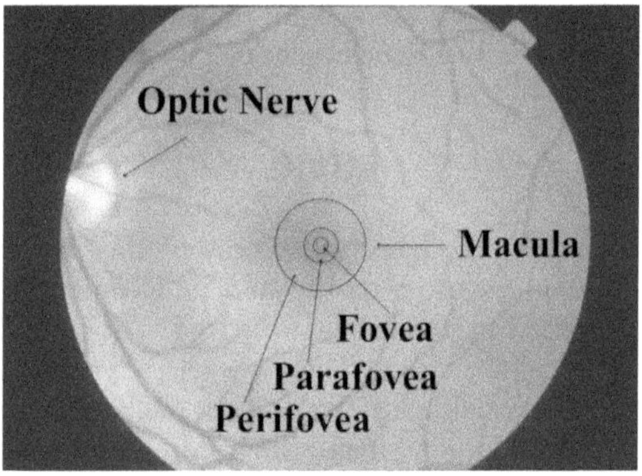

FIGURE 2.7 *The fovea, parafovea, and perifovea.* Source: *National Eye Institute, National Institutes of Health. Modified by the authors.*

(farsightedness) the focus is affected and is too far or too close, resulting in reduced vision. The fovea is lined with cone photoreceptor cells, see Cones section.

We further distinguish two concentric areas around the fovea proper: the parafovea and the periphery of the fovea (perifovea). The visual acuity progressively diminishes as we move away from the fovea. This will become particularly significant in the reading applications of eye tracking (see Chapter 5). Figure 2.7 provides a schematic representation of the fovea and its regions. Note that the image is not drawn to scale: the fovea has a diameter of 1.5 millimeters and the perifovea's diameter is 5.5 millimeters. This image of the fundus of the eye (i.e., the inner lining of the back of the eye) is taken with a special camera, essentially a low-powered microscope, specifically designed to take pictures of the fundus.[1] It reveals the retina (the background), the optic nerve and the macula. Within the macula, we have highlighted the fovea, the parafovea, and the perifovea. Note also the capillaries providing blood flow to the retina. As a fun fact, let's add that the fundus is responsible for the red-eye effect when taking flash photography: the light of the flash is reflected by the fundus, resulting in the infamous red glow.

Why Is the Image Reaching the Fovea Inverted?

An interesting detail of vision is that the retina receives an upside-down image of the world. This is due to the optics of lenses. A lens is usually glass or a similar transparent material. Because of the curvature of the lens, it alters the direction of the rays of light that traverse it. In a converging lens (called a convex lens) the rays are deviated in such a way that they converge on one point, called the focus. This is why you can light a fire using a magnifying glass, which makes the sun rays converge on one point, which gets very hot. Imagine putting a projection screen at the focal point and parallel to the lens. What you would see is an image of the object in front of the lens, but the image would be upside down. The size of the image will depend by how close to the lens your screen is placed. Figure 2.3 should clarify the reason for the inversion of the image. Notice how the rays starting from pointed part of the arrow arrive from the top of the image, pass through the focus, and arrive at the bottom of the projected image and vice versa the rays that start at the bottom arrive at the top, after passing through the focus as well.

However, despite the fact that the image which reaches the retina and specifically the fovea is upside down, there is no "flipping" of the image in the brain. The retina is not a passive screen on which the image is projected; it is, as we will see in the following, an array of photoreceptor cells that are activated by light and that send a "message" to the brain. In other words, the message is sent already flipped (i.e., it reaches the brain right side up).

Optic Nerve

The optic nerve carries the entire output of the retina to the brain. The visual information is acquired and processed in the retina and is sent unaltered along the optic nerve to the vision centers of the brain. Anatomically, the optic nerve connects the retina to the optic chiasm in the brain. The optic chiasm is an area in the brain in which the nerves coming from each eye cross in a "X" shape so that the left eye goes to the right side of the brain and vice versa. However, some of the fibers from each eye go to the same side of the brain so that each side of the brain receives information from both eyes. This allows the perception of depth. It has been claimed that the evolution of the optic chiasm is a turning point in the evolution of hand-to-eye coordination (Larsson, 2013). The optic chiasm is clearly visible in Figure 2.1 as the central x-shaped structure. We will examine the optic nerve more in depth in Chapter 3.

Photoreceptor Cells

Photoreceptor cells are cells that are specialized in converting light into biological signals and specifically electrical signals at the membrane (external layer) of the cell. These are also called membrane voltage or membrane potential. As we have seen, there are three kinds of cells: the rods, the cones, and the **ganglia.**

Rods

Rods are elongated, as the name will remind you. Rods work best in dim light, as they are extremely sensitive to light, so much so that a single photon (you can think of photons as particles of light) can trigger them. In low-light conditions all vision is reduced to rods. Rods are most sensitive to movement and shapes. Since rods do not detect color, it is difficult to guess the color of objects at night, but one can still see shapes and movement. Estimates of the number of rods in the human eye range from 90 to 130 million. Rods are more concentrated in the periphery of the retina and diminish in density as they get nearer to the fovea.

Rods are so sensitive to low amounts of light that the human eye can detect as few as six individual photons at a time! In other words, six single rods activated are enough for us to perceive light. Rods can perceive up to 10,000 photons per second after which they saturate; that is, they stop working, because they are overwhelmed by the light and the cones step in.

Cones

Cones are roughly shaped like a Christmas tree. They work best in bright light and are less sensitive to light than rods: this allows them to perceive as many as 10 billion photons per second (full sunlight on a snow field). They are mostly responsible for color vision. There are three different kinds of cones, sensitive to different wavelengths of light: blue (low), green (medium), and red (high). Estimates of the number of cones range between 4.5 and 6 million. Rods and cones are mixed in the retina, except for the fovea, which is an area of the retina of about 0.2/0.3 mm diameter, which is

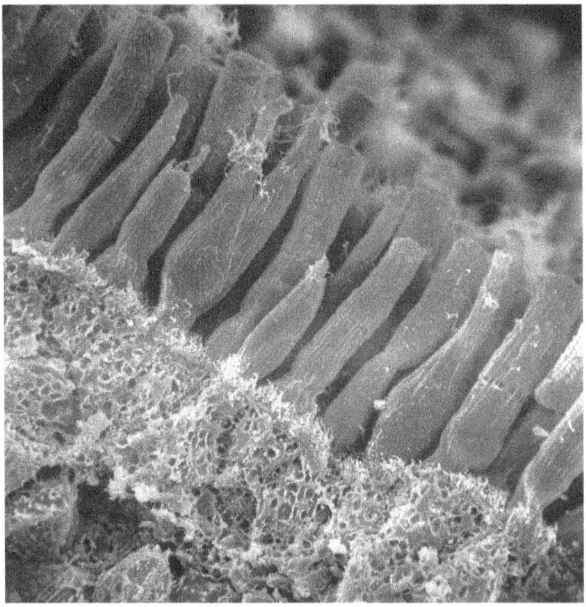

FIGURE 2.8 *A detail of the retina, with cones and rods clearly visible.* Source: *Science photo, reproduced with permission.*

without rods (see Fovea Centralis section). In Figure 2.8, you can clearly see rods and a few cones cells in the front.

Photosensitive Ganglia

The photosensitive ganglion cells (ganglia) have two main functions:

1. Regulate the circadian rhythm. The circadian rhythm is the alternation of sleep and wakefulness in the body. The term comes from Latin *circa diem* as the circadian rhythm matches roughly a day. The circadian rhythm regulates the impulse to go to sleep (sleepiness).
2. Contribute to the regulation of the pupil: As we saw the pupil is sensitive to light and contracts under the right light conditions. The ganglion cells detect the intensity of light and trigger the contracting. They essentially act as a brightness detector.

There are between 2 and 3 million ganglion cells but a small percentage (between 1 and 3 percent) are photosensitive. The optic nerve is made of the axons (nerve fibers) of the ganglion cells. The **blind spot** is an area of the retina in which ganglion cells are gathered and leave the eye as the optic nerve. As the term "blind spot" indicates, there are no photoreceptor cells in this area and hence no vision takes place in the area. It merely transfers the information to the brain, via the optic nerve.

Conclusion

We can summarize the path that a light ray takes in the eye as follows: light is reflected off an object in the world. The light ray enters the eye by traversing the conjunctiva, the cornea, the aqueous humor, the iris (via the pupil), the lens, where it is focused, and the vitreous humor. It finally reaches the retina and within it the macula and the fovea. At this point it is transformed into electrical impulses and biochemical information, which is gathered in the retina and funneled via the optic nerve to the brain. We turn to this second phase of the process in the next section.

From Light to Nervous Impulses to the Brain

As we saw earlier, a ray of light enters the eye and hits the retina. Light is made of photons. In the retina the rods and the cones absorb the photons and through an electrochemical process, discussed in Chapter 3, transform the light into nerve impulses. The impulses travel along the optic nerve. Since we have two eyes, essentially a second image travels along from the other eye to the brain in parallel. As we already saw, in the optic chiasm, some of the visual information from the left eye goes to the right side of the brain, but some of it also goes to the left side, and vice versa. This allows stereoscopic vision, which gives us sense of depth (e.g., how far away objects are). Finally, the electrical impulses arrive at the visual cortex, a region of the brain in the back where vision is processed. We will examine the neurology of vision in the next chapter.

Self-Study Questions

The following questions pertain to the contents of this chapter. You should be able to answer them without looking up the answer in the book.

1. What is the difference between anatomy and physiology?
2. Describe the functioning of the eye.
3. What is the fovea? Where is it located?
4. What are the three kinds of photoreceptor cells?
5. What is the pupillary light reflex?

6. What is the blind spot?
7. Explain why the image on the retina is inverted.

Words to Know

sclera	limbus	cornea	conjunctiva	iris
pupil	pupillary light reflex	anisocoria	pupillary hippus	lens
aqueous humor	vitreous humor	retina	fovea	photoreceptor
rods	cones	ganglia	blind spot	latency

Further Reading

Our treatment of the anatomy and physiology of the eye is dependent primarily on Records (1979) and chapter 2 of Duchowski (2003; 3rd ed., 2017).

CHAPTER 3

The Visual System in the Brain

This chapter describes the neurological basis of vision. It is the continuation of the previous chapter, as it starts from the image on the foveal area of the retina and follows the path of the nervous signals all the way to the **visual cortex** in the brain. Figure 3.1 gives a schematic idea of the organization of the visual pathways in the brain. You can imagine the information gathered in the back of the retina traveling down the optic pathway, splitting in half at the optic chiasm (note how the darker/lighter[1] lines go to *both* sides of the cortex), reaching the lateral geniculate nuclei, spreading out in the **optic radiation**, and eventually reaching the visual cortex. So, in a sense, this chapter is a "map" of the visual pathways.

Vision is very concretely embodied, in a way that other systems, such as language, are not. Quite literally, we see what our visual receptor cells allow us to see. With language, unless you are engaging in neurolinguistics, you don't need to worry about which parts of the brain support language. With vision, we don't have that luxury of abstraction. So, it is necessary to understand *how* we see what we see. It should go without saying that what follows is a very rapid and schematic presentation. Our goal is to give the reader a "feel" for how the visual system in the brain works. For more in-depth treatments, refer to the further reading section.

Figure 3.2 may be useful in orienting yourself in the brain. Each hemisphere is divided into four lobes, and areas are described in relation to these lobes. So, for example, we may say that the visual cortex is located in the posterior occipital lobe (meaning, the back part of the back of the brain).

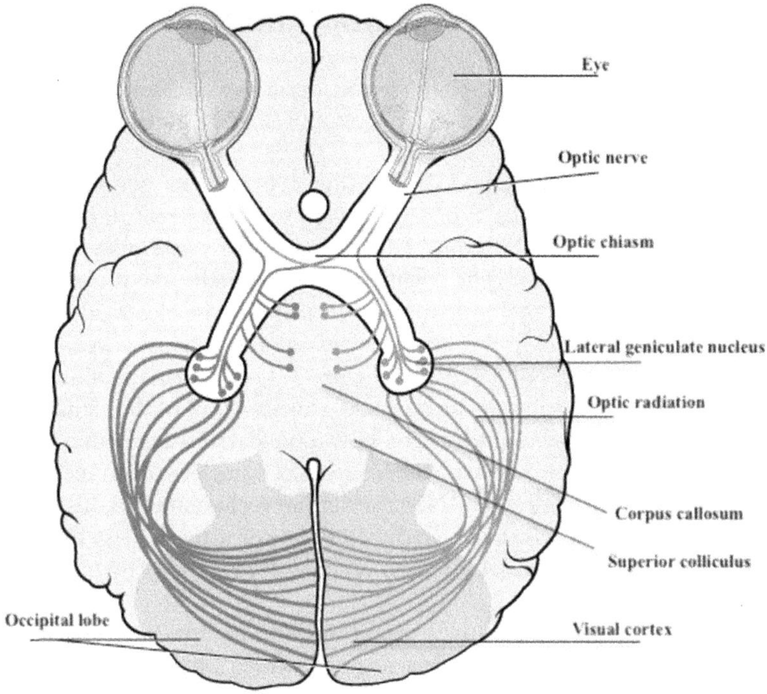

FIGURE 3.1 *Optic pathways.*

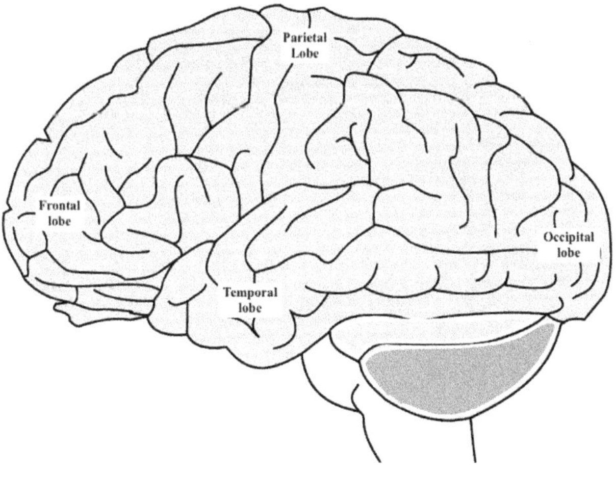

FIGURE 3.2 *The lobes in the brain.*

Physiology of the Visual System

Optic Nerve, Chiasm, and Tract

As we saw in the previous chapter, the optic nerve connects the eyes (specifically, the neural output of the retina) to the brain, via the optic chiasm.

The optic nerve is composed of ganglion neuron cells (see Chapter 2), which have long axons. The axons are coated with **glial cells**, largely made of a fatty substance called myelin, which insulate the nerve cells from each other. The coating of glial cells is called the myelin sheath of the neuron. The overall function of the optic nerve is to transfer the electrical impulses from the photoreceptors to the striate cortex (aka, the visual cortex). There are many different types of nerve cells. Figure 3.3 is a schematic illustration.

In order to understand the nature of the transmission of the signal along the optic nerve, we need to understand how a typical nerve cell functions, in this respect. As Figure 3.3 shows, a nerve cell has a nucleus and a membrane which envelops the entire cell. Nerve cells have elongated axons. Axons can vary in size from less than a millimeter to over a meter (less than one-sixteenth of an inch to over 40 inches). The dendrites are a few millimeters in size.

Each cell in isolation can only transmit a yes/no pattern (i.e., whether the cell is firing or not; see later in this section), which obviously is a limited amount of information and so would not be that useful. However, cells are interconnected and by working together, they can convey much more complex information. Let us consider two connected cells: the end part of the axons (axons terminals) of the first cell will be in contact with, but not attached to, the dendrites around the body of the second cell. Each cell will

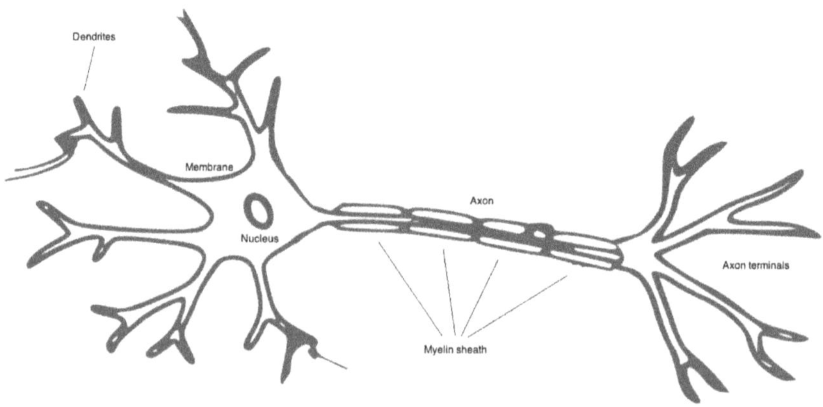

FIGURE 3.3 *Schematic representation of a nerve cell.*

be in contact with hundreds of other nearby cells, of course. So we should not think of this as a string but more of a network. This will become very significant when we consider how clusters of cells fire (see The Electrical Impulse in the Axons section).

As we saw, the axons and the dendrites do not quite "touch" each other. There is a gap between them, the **synaptic cleft**, which is admittedly very small (10–20 nanometers).[2] Figure 3.4 is a schematic representation of the synaptic cleft. The cell from which the signal comes is called **presynaptic** and the one receiving the signal is called **postsynaptic**.

So, if the dendrites and the axon terminals do not quite touch, how does the signal travel across the synaptic cleft? The answer is by **chemical transmission** (see Chemical Transmission section). The following section considers the transmission of impulses in nerve cells. There are two ways for a signal to travel: an electrical transmission and a chemical transmission. Both are usually used by a cell.

The Electrical Impulse in the Axons

The process is rather complex and we can only sketch it very crudely. The main point is that there are no "electrical impulses" that travel down the optic pathways. It's just the signal that travels. A good analogy would be two children holding a jumping rope and making a "wave" travel along its

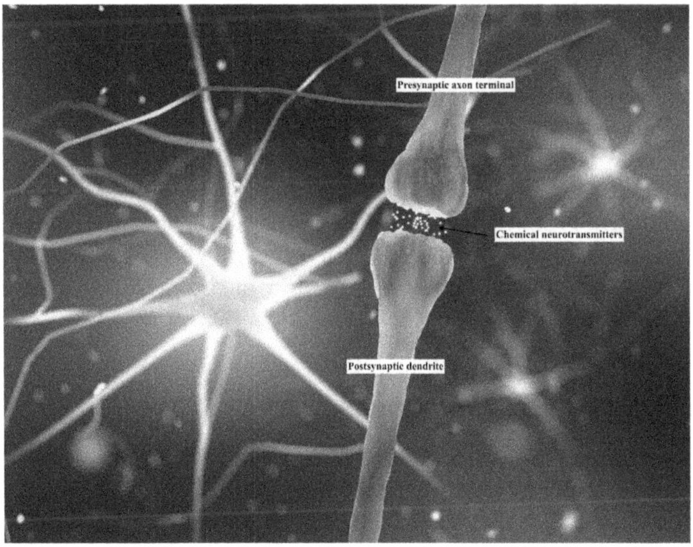

FIGURE 3.4 *Schematic representation of a synaptic cleft. In the background, a cell with axons protruding from the nucleus.* Source: *Science photo, reproduced with permission. Modified by the authors.*

length: clearly the fibers of the rope do not move, but the motion of the rope up and down "travels" along the rope.

In the case of nerve cells, the "wave" that travels along the axon is a wave of opening of receptors along the body of the axon. Basically by opening and closing these receptors (you can think of them as little "doors" that let in and out ionized[3] atoms of sodium and potassium) the cell membrane acquires an electrical charge. When the charge reaches a certain threshold, it fires and thus the charge travels along the axon, until it reaches the axon terminals and it is then transferred to the postsynaptic cell. The speed of the impulse along the axon is between 0.1 and 10 meters per second (4 inches per second to 30 feet per second). After it has fired, the part of the axon is unable to fire again until it recharges, so to speak. The rate of firing of the impulse is between 100 and 200 impulses per second, but may go up to 800 per second.

Chemical Transmission

As we saw, the impulse travels along the axon but *not* across the synaptic cleft. The firing of the presynaptic cell triggers the production of chemicals (**neurotransmitters**) which cross the synaptic cleft (see Figure 3.4). Among the **neurotransmitters**, the most common is glutamate. Other neurotransmitters you may have heard of include dopamine, adrenaline (epinephrine), and noradrenaline (norepinephrine).

Role of the Glial Cells

Glial cells are largely made up of myelin. They coat the axons. Myelin is primarily fatty and has a white color; it is what gives the brain and spinal cord's white matter its distinctive color. Conversely, gray matter is made up of the cell bodies, dendrites, and synapses but may also contain some axons coated with myelin. Glial cells support, feed, and protect the cells from pathogens. From the point of view of signal transmission, the function of the myelin-rich glial cells is also to speed up the transmission of the activation along the axon and play a role in the connections between the nerves (synapses).

Convergence and Divergence

As we saw earlier, virtually all cells are connected to more than one cell; that is, they receive input from many other cells. This is called **convergence**. The opposite (i.e., the fact that each cell sends its output to hundreds of other cells) is called **divergence**. Some of these inputs are excitatory (i.e., they tend to make the cell fire), but some are inhibitory (i.e., they tend *not* to make the cell fire). So there is no one-to-one signal. Nerve cells are not conduits; they elaborate (either reinforce or tamp down) the signal. In this complex system, in which most cells are connected to numerous other cells and may receive different and even contradictory messages, how does any given cell "decide" whether to fire or not? There is a sort of "majority rule" effect: the cell integrates (adds up, so to speak) all the excitatory or inhibitory effects and if a given threshold is met, it fires. So, in conclusion, as we already observed, the function of the neurons is not just to transmit the impulse but also to integrate all the impulses coming from the cells they are in contact with and thus determine whether to fire or not.

The Optic Chiasm

The optic chiasm is so called because it has the shape of the Greek letter *Chi* (X) and, as we saw, it is clearly identifiable in the schematic representation in Figures 2.1 and 3.1. The technical term for the crossing over of the nerve is **decussation.**

As can be seen from the schematic representation in Figure 3.1, part of the nerve pathways (darker color) coming from the left eye ends up in the right cortex and part on the left, and likewise part of the pathways from the right eye (lighter color) ends up in the right and left cortex so that each side of the visual cortex receives information from *both* eyes. As we saw, it is believed that this feature of the vision system is crucial in the development of hand-to-eye coordination. However, even more importantly it allows **stereopsis** (binocular vision) and thus the perception of depth and three-dimensionality. Essentially, each eye receives a slightly different image and by comparing them, the perception of depth emerges.

The Role of the Corpus Callosum

The **corpus callosum** (Figure 3.5) is a bundle of nerves centrally located in the brain, which joins the two halves (hemispheres) of the brain.

It allows both hemispheres to communicate with each other (exchange information). In particular, it combines information from both halves of the visual field (see Figure 3.1) and connects the visual information to the language centers of the brain, which is crucial to be able to identify what we are looking at.

The Lateral Geniculate Nucleus, Optic Radiation, and Visual Cortex

After crossing the optic chiasm, the optic nerve "terminates" in the lateral geniculate nuclei (there is one for each hemisphere). See Figure 3.1 for an overall view. By "terminate" we mean that the fibers of the optic nerve

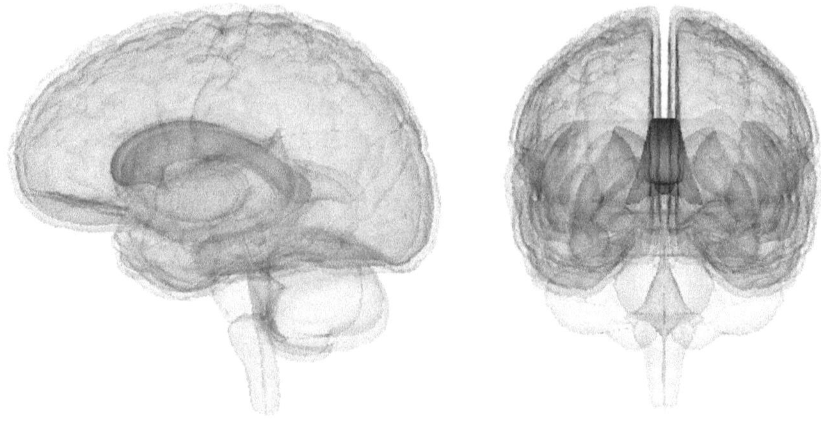

FIGURE 3.5 *An image of the corpus callosum using X-ray imaging.* Source: *BodyParts3D, © The Database Center for Life Science licensed under CC Attribution-Share Alike 2.1 Japan.*

fan out in the midbrain area and eventually reach the visual cortex in the occipital, dorsal lobe (the back of the brain). In the rest of this section we examine each of the three major components.

Before we do so, however, we must establish a rather counterintuitive fact: the visual pathways are **topographically** organized. This means that cells located near to one another are connected, but also that these connections, based on the "topography" (think, geography) of the cells in one region of the visual pathway, are preserved in the following regions. So, if two cells are associated in the retina, they will also be associated in the **lateral geniculate nucleus** (LGN) and in the visual cortex. This is particularly counterintuitive when you realize that in the optic nerve they "become almost completely scrambled" (Hubel, 1995, p. 61) and then they are disentangled again in the lateral geniculate and optic radiations. However, the **topographical organization** does not imply a one-to-one correspondence: for example, the foveal region is represented by a much larger region in the visual cortex than the peripheral region of the retina.

The Lateral Geniculate Nucleus

The LGN is a structure in the midbrain, where the optic tract terminates. It consists of six layers of cells: the first two are **magnocellular** (M cells) and layers 3–6 consist of **parvocellular cells** (P cells).[4] As the Latin names indicate, the magnocellular cells are larger, whereas the parvocellular ones are smaller. The M cells receive the information from the rods, whereas the P cells receive it from the cones. There are other even smaller cells (koniocellular). Each type of cell receives information directly (topographically) from corresponding ganglion cells in the retina. Much like rods and cones react to different degrees of light and wavelengths, the M and P cells react to different stimuli. So, the pathway from the ganglion cells in the retina to the M cells in the LGN is called the M pathway and correspondingly we find a P pathway as well. The M pathway is good at motion detection and light/dark detection, for example, whereas the P pathway is good for color and high resolution. The pregeniculate nucleus (PGN) is a small part of the LGN, although it is separated from it by fibers of the optic tract. It is associated with saccadic movements.

The Optic Radiation

From the LGN, optic radiations (i.e., axons of neurons in the LGN) fan out toward the visual cortex. Remarkably, the topographical organization of the nerve fibers remains constant until it reaches the visual cortex. However, we must not forget that cells receive information from adjacent cells as well (convergence; see Optic Nerve, Chiasm, and Tract section) and thus these mappings get less direct as we move away from the retina. To put it differently, the brain elaborates the information. Of course, the main locus of visual information processing is the primary visual cortex.

The Visual Cortex

The visual cortex (aka, striate cortex, due to a line on it that is visible to the naked eye) is the primary center of the brain for visual processing. It is located in the back of the brain (see Figure 3.6). There are in fact several

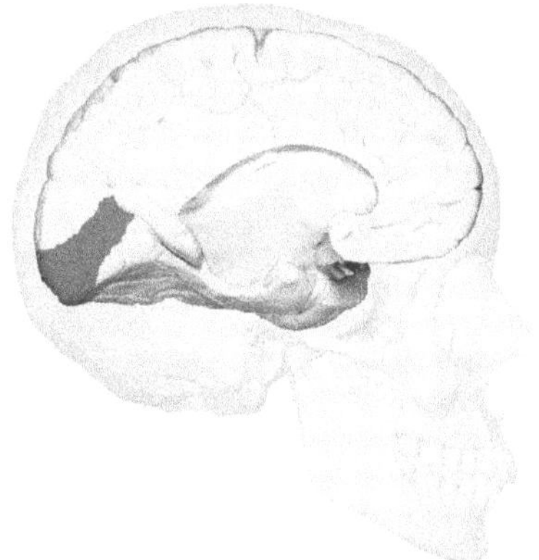

FIGURE 3.6 *An image of the visual cortex using X-ray imaging.* Source: *BodyParts3D, © The Database Center for Life Science licensed under CC Attribution-Share Alike 2.1 Japan.*

areas that form the visual cortex; they are labeled V1 (the primary visual cortex, also labeled Brodmann[5] area 17) and V2 through V5. The areas have various specializations: area V1 detects orientation (up, down, etc.); areas V2–V4 form/shape and color processing; area V5 processes motion (it is also called the middle temporal area, MT).

The visual cortex contains a variety of cells that are sensitive to very specific kinds of visual information. For example, individual cells may be sensitive to movement but barely fire or not at all in the presence of a static stimulus. Other cells are color, shape, or form sensitive; others are sensitive to borders. Others still are sensitive to motion, but only in a given orientation: they may fire at a movement at 45 degree but not at one at 90 degree. One of the most counterintuitive results of the studies of vision is that when we see a uniformly colored square, for example, we actually only *see* the borders, the edges of the square. The central part is interpolated by the brain, but the actual visual cells do not fire for it.

The cells are organized in layers. A good metaphor is a blanket, folded over three or four times and then scrunched up. The folds correspond to the ridges of the cortical matter (gyri) and the gaps to the fissures between them (sulci). See Figure 3.7. Figure 3.8 presents a view of the brain with some notable gyri labeled.

The specialization of the cells is so extreme that the phenomenon known as Troxler effect (see Chapter 1) emerges, since there are no cells dedicated to firing for stationary objects. In fact, if one stabilizes an image on the retina, after about one second the scene disappears and the field of vision is blank. As we saw, the slightest movement restarts the vision process. It is, as Hubel (1995, p. 81) put it, “as if the visual system, after going to the trouble to make movement a powerful stimulus—wiring up cells so as to be insensitive to stationary objects—had to invent microsaccades to make stationary objects visible.”[6] We will discuss microsaccades in more detail in Chapter 4.

The reader should not come to the conclusion that the visual cortex is the endpoint of visual processing. Rather, it is the beginning of a series of processes. Each area of the visual cortex is connected to other areas and

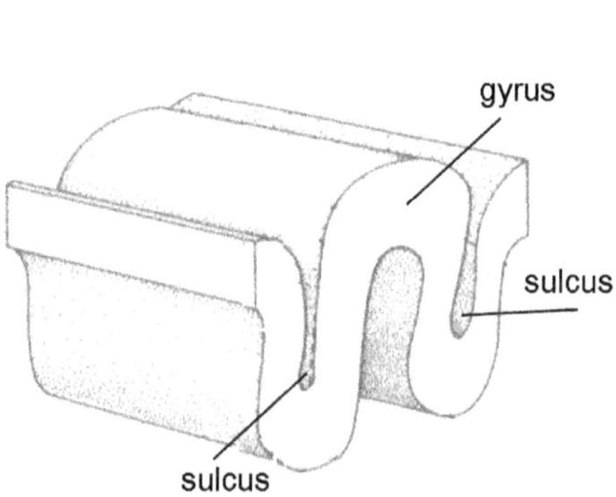

FIGURE 3.7 *Sulci and gyrus. Illustration by Albert Kok. Wikimedia Commons.*

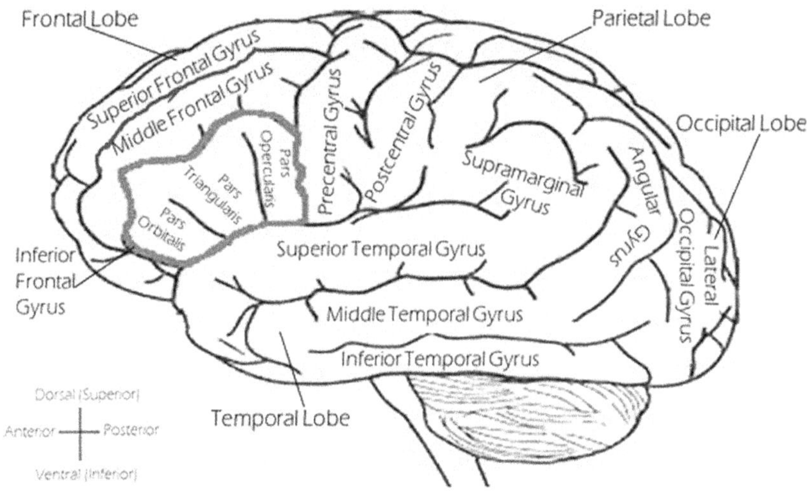

FIGURE 3.8 *Some notable gyri.* Source: *Pixabay.*

thus quite clearly the processing of the visual input continues in those areas, perhaps at a higher level of abstraction.

Role of Other Brain Structures in Attention and Saccades

This section briefly reviews other neural structures in the brain that have a significant impact on vision and eye movement and have not been considered so far in our exposition.

The Intraparietal Sulcus

The **intraparietal sulcus** is located in the parietal lobe (see Figure 3.9) in the dorsal (superior) posterior (back) part of the brain.

The intraparietal sulcus is involved in visual attention, saccades, perception of depth in stereopsis, and hand-to-eye coordination.

The Superior Colliculus and Frontal Eye Field

The **superior colliculus** is located in the midbrain, right above the pons and near the cerebellum. See Figure 3.10.

The superior colliculus is largely responsible for the targeting of objects of interest by the saccades. Essentially, it receives input from the retina and the visual cortex and determines the difference between where the fovea is directed and where it needs to be centered on the object of interest. Another region of the brain, the **frontal eye field**, located in the frontal cortex actually

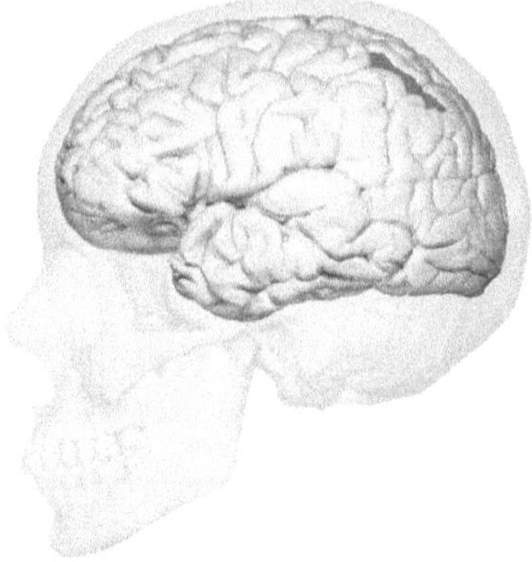

FIGURE 3.9 *The location of the intraparietal sulcus.* Source: *BodyParts3D, © The Database Center for Life Science licensed under CC Attribution-Share Alike 2.1 Japan.*

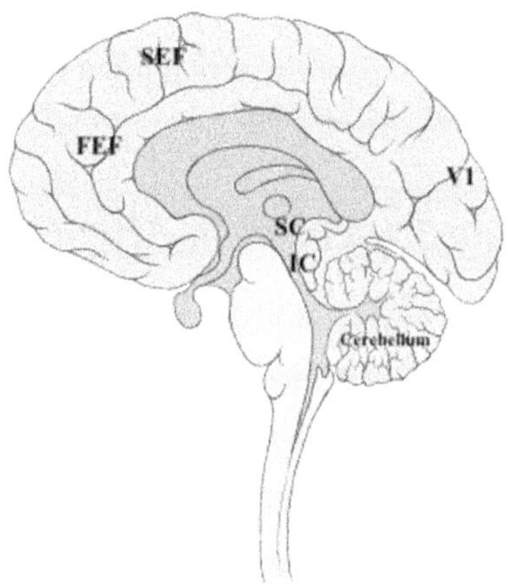

FIGURE 3.10 *The location of the frontal eye field, superior and inferior colliculi, and visual cortex.* Source: *Patrick J. Lynch, Wikimedia Commons. Modified by the authors. FEF = frontal eye field; SC = superior colliculus; IC = inferior colliculus; V1 = visual cortex; SEF = supplementary eye field.*

does the moving of the eyes (Figure 3.10). The frontal eye field is involved in attention and control of the eyes. It has access to the muscles that move the eye balls and thus governs the actual movement, both voluntary and involuntary (saccades).

Both the superior colliculus and the frontal eye fields are also believed to contain a population of cells that function as a saliency map (see Chapter 9), that is, as a representation of the most salient areas of the visual scene, which is then used to determine the target of the next saccade.

Lateralization, Hemispheric Specialization

The fact that the overall symmetry of the brain[7] is belied by hemispheric specialization or **lateralization**, that is, the fact that some tasks are better handled by one hemisphere rather than the other, should come as no surprise to linguists, who should be familiar with the Broca's and Wernicke's areas, which were the earliest structures connected with language processing and are found in the left hemisphere. More recent work has confirmed the presence of several processing asymmetries. For example, the left hemisphere is associated with syntax, whereas the right hemisphere is associated with prosody. Lesions in the corpus callosum, which prevent communication between the two hemispheres (see Optic Nerve, Chiasm, and Tract section), have been shown to interfere in normal processing (Friederici, 2011). Thus, for example, an incorrectly placed intonational phrase boundary elicits an N400 ERP[8] in people with intact corpus callosum but not in patients with lesions (Friederici, 2011, p. 1384).

As far as the visual system goes, lateralization is shown in the perception of faces, for which the right hemisphere is specialized, as can be seen, for example, from work with "chimeric faces," that is, faces in which the left or right side has been replaced by a copy of the other side, which show that right-right faces resemble the original more than left-left ones. Another source of evidence of lateralization is studies on people with prosopagnosia (i.e., incapacity to recognize faces) caused by lesions in the right hemisphere. Recent work with brain imaging shows a predominance of the right hemisphere (while also showing clearly that both hemispheres are involved). Another set of phenomena that demonstrates lateralization is related to the perception of gaze direction, which is also specialized in the right hemisphere, as shown by experiments in which gaze presented in the left visual field is processed better—recall that the left visual field is processed in the right hemisphere (Brancucci et al., 2009).

Conclusion

The discussion of the processing of visual information in the brain concludes our discussion of the visual pathways of the nervous system. It also completes the discussion of the eye and vision in general. We are thus able to understand how the light photons entering the eye are transformed into the retina in electric/chemical signals that travel along the axons of the optic nerve and eventually reach the visual cortex and are then distributed to the brain for conceptual or motor processing. The chapter should have disposed of two naive metaphors of vision: the first, that an analog image of reality is somehow projected in the brain: vision is very much digital, not in the sense of pixelation, which is a different process actually fairly close to analog, but in the sense that the neurons transmit binary, yes/no information, which is processed by the brain into a representation of reality that we "see." The second, that electrical impulses "travel" down the neurons from the eye to the brain. There are no continuous nerve connections, as even for the longer axons there are always synaptic connections and moreover the axons do not convey electrical charges like a copper wire, and finally, the synaptic connections are mediated by chemical neurotransmitters. So, again, there is no "direct" transmission of images or signals. We see, literally, through our brain.

Self-Study Questions

The following questions pertain to the contents of this chapter. You should be able to answer them without looking up the answer in the book.

1. Describe the optic pathways starting from the retina and ending in the visual cortex.
2. What is the optic chiasm?
3. Explain topological mapping.
4. What evidence is there for lateralization?
5. What is integration and what role does it play in vision?
6. What is the difference between the following cell types: rods, cones, cellular, parvocellular, neurons, and glial cells?

Words to Know

synaptic cleft	presynaptic cell	postsynaptic cell
chemical transmission	neurotransmitters	convergence
divergence	decussation	stereopsis
corpus callosum	topographical organization	magnocellular
parvocellular	intraparietal sulcus	superior colliculus
frontal eye field	visual cortex	optic radiation
glial cells	axons	lateral geniculate nucleus
lateralization		

Further Reading and Sources

The chapter is based largely on Hubel (1995), Snowden et al. (2012), and Schiller and Tehovnik (2015). They are all accessible, relatively speaking. There are many online resources that are well worth the time. Biga et al. (2019) chapter on vision in their *Anatomy and Physiology* textbook is a good place to start, but it is aimed at medical students and thus may be a little challenging: `https: //open.oregonstate.education/aandp/chapter/15-5-vision/`.

CHAPTER 4

Eye Tracking Basics

Eye tracking consists of the measurement of eye movements, primarily **saccades** and **fixations**. When you were first introduced to phonetics and phonology, in order to be able to understand what a fricative is, for example, you were introduced to the phonatory organs (lungs, tongue, lips, etc.). Here you need to be introduced to the kinds of eye movements that are measured and how they are measured. This is what this chapter will do. We will start by examining the types of eye movements and then we will move on to the types of measurements used in eye tracking and the visualizations used to display them.

Eye Movements

The eyes can move in the eye sockets due to the presence of six muscles (the extraocular muscles, so called because they are external to the eye itself), illustrated in Figure 4.1. There are three pairs of muscles; when one muscle in a pair is activated, the other one is inhibited (not activated) and thus the eyes can move up or down, left and right, and clockwise and counterclockwise (**torsional movements**; see Figure 4.2), in a 3D space.

If you are having trouble visualizing the three-dimensional movements, try this: make a fist and hold your arm straight in front of you. Imagine that your fist is an eyeball. If you move your fist left to right and right to left, you have your first dimension (horizontal movement). Now move your fist up and down, while keeping the arm still: that's your second dimension (vertical movement). Finally, turn your fist as if to turn a key, clockwise and counterclockwise: that's your third dimension (torsional movement).

Eye movements are usually measured in terms of angles. A circle is divided in 360 degrees (or 360°) and 90° is a right angle. The human visual field is about 110°. The usefulness of measuring in angles is that the distance from

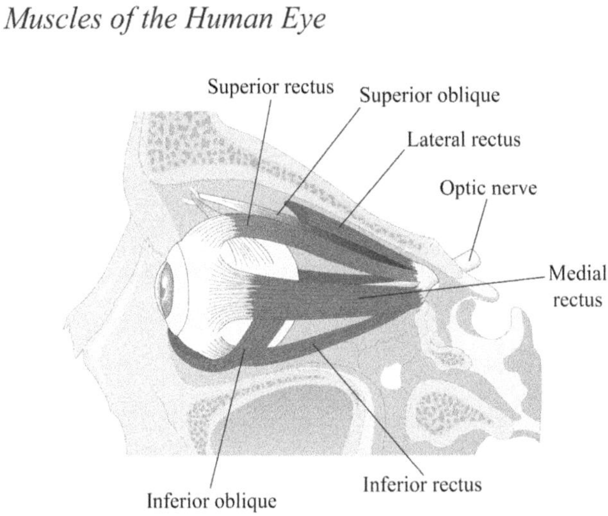

FIGURE 4.1 *The muscles of the eye.* Source: *Science photo, reproduced and modified with permission.*

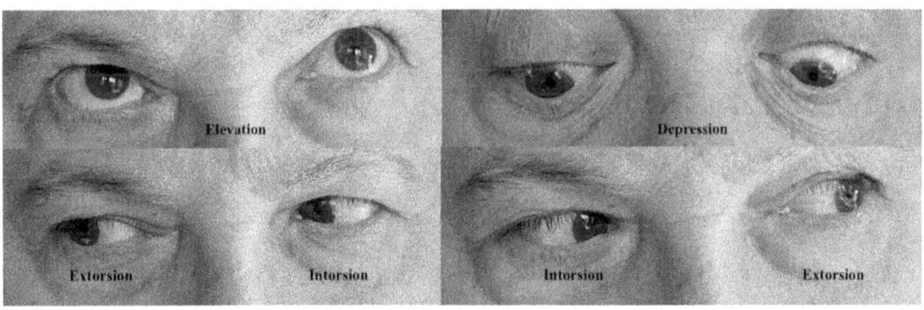

FIGURE 4.2 *Vertical and torsional movements.*

the object in focus is not relevant: if one moves one's eye by 5° while looking at something close, the focus will move only by a few inches, whereas if one is looking at the moon, the same 5° movement will involve a much greater movement of the focus. The size of saccades is usually measured in degrees as is the size of the fovea, parafovea, and periphery (on the fovea, see Chapter 2). Measured in degrees, foveal vision is about 1.5° of the visual field. A common rule of thumb (pun intended) is that the foveal area of maximum visual acuity is the size of your thumbnail at arm's length.

Saccades

The purpose of saccades is to orient high-resolution foveal vision toward targets of interest. Saccades are ballistic movements of the eye, that is, fast, broad, and not very precise. They consist of five phases (see Figure 4.3).

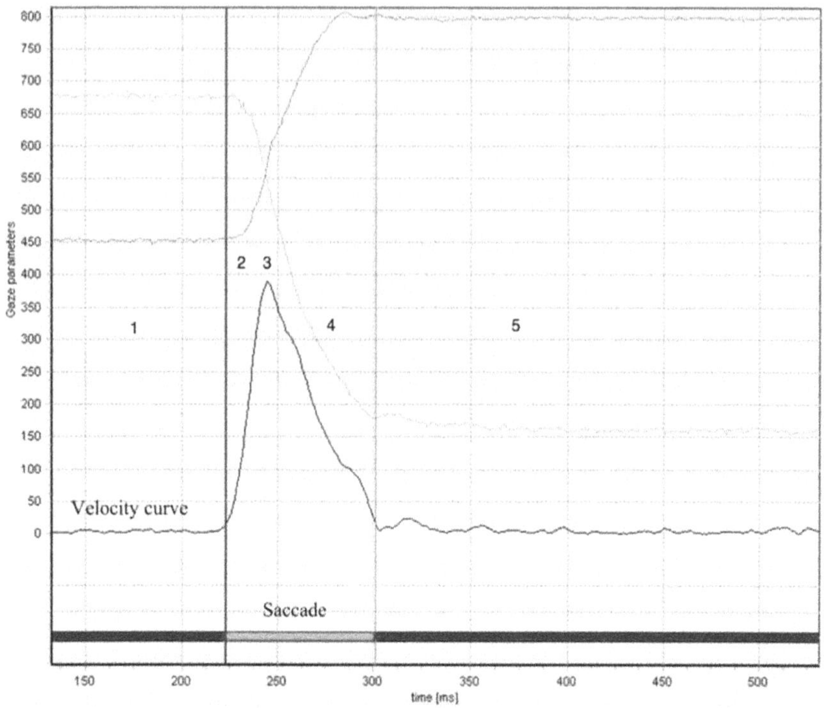

FIGURE 4.3 *The five phases of a saccade (velocity).* Source: *Cui et al. (2014); modified by the authors.*

1. A stable phase (essentially, a fixation; see Fixations section)
2. Very quick acceleration
3. Maximum speed (lasting from 10/20 ms to 100 ms)
4. Very quick deceleration (slightly more gradual)
5. A stable phase

Once they have been initiated they can no longer be corrected. In this sense they are similar to a jump: once you have left the ground you can no longer control your direction. Saccades involve both eyes, under normal circumstances. They are controlled by the frontal eye field and the superior colliculus (see Chapter 3). They are mostly linear (since that is the most efficient way to move from point a to point b) but they can be curved under special circumstances, such as being affected by a distractor. For example, a light or moving object that appears during the saccade. The biggest distance of a saccade is 90°, but for any saccade larger than 15° or 20°, the saccade is accompanied by head movement. However, generally saccades are much smaller; for example, in reading saccades tend to be between 1.5° and 3°.

Once a saccade has been initiated, there is a period in which the eye is unable to perform another saccade. This is called **latency.** The duration of latency varies quite significantly between 100 ms and 1,000 ms. Voluntary saccades require longer latency. Within a **scan path** (see Gaze Paths section) latency is defined as the intersaccadic interval (i.e., the time between successive saccades). In experimental settings, latency is generally defined as the time elapsed between the stimulus and the onset of the movement.

The two-step paradigm (Becker & Jürgens, 1979) consists of asking the subject to follow a dot. In some cases, the dot moves to a second location. If the movement occurs close to the execution of the saccade, the saccade is directed to the original position of the dot, whereas if the movement occurs longer before the saccade, it is directed to the second location. This shows that there is a "point of no return" after a saccade has been planned, called saccadic **dead time,** and that if the target is moved too close to the dead time, the viewer has already committed to the saccade and is unable to correct. Studies using the two-step paradigm have shown that there needs to be at least 70/80 ms between saccades for the eye to be able to plan another landing point.

The target of the saccade is selected either externally (i.e., a new stimulus) or determined through peripheric vision. We will discuss this in more detail in Chapter 5. Saccades can also be exploratory, for example, when the viewer scans the scene for items of interest. Exploratory saccades take place at a higher rate than, for example, saccades during reading (Otero-Millan et al., 2013). Saccades during search are broader than during memorization of a scene. Corrective saccades are short saccades that "correct" the landing point of a larger saccade. They usually occur very quickly after the first saccade and are not planned.

During saccades we are for all intents and purposes blind: the movement of the eyes is too fast for the retina to be able to process the images. This is called **saccadic suppression.** How are we unaware of this fact? Essentially, the brain compensates and as a result we have no perception (or only a marginal one) of saccades. The superior colliculus (see Chapter 3) which plans the saccades also sends a copy of the planned saccade to the frontal eye field (see Chapter 3), which then sends it back to the cortex as a "warning of an impending eye movement" (White & Munoz, 2011, p. 199). This then allows the brain to compensate for the saccade and yields the uninterrupted perception of vision we enjoy.

Fixations

Fixations are defined as relatively static periods during vision in which the foveal vision is fixed onto a small area. It is very important to recall that the eyes are never completely static and that in fact if they are made to be still, they lose the capacity to see (Troxler effect). However, compared to the large-scale movements of saccades, or to tracking motions, for example, fixations are relatively still. Thus, a fixation is an abstraction, because in natural settings the eyes are never still. Normally, the eye-tracking equipment you will use will have default settings that determine the threshold of movement under which a series of movements is considered to be a fixation.

It is necessary to consider how an eye tracker works, concretely. Using some method (see Chapter 1) of tracking the position of the fovea, the eye tracker produces a series of data points, essentially time-stamped (x,y) coordinates in the field of vision. Imagine playing battleship with a moving target: at time 1, your target might be in location (a, 5), at time 2 in location (a, 6), and at time 3 in location (b,7). Each set of coordinates is a data point, that is, one sample of the position of the fovea.

The sampling frequency of eye trackers varies quite significantly from as low as 30 Hz to as high as 2,000 Hz. Hertz (Hz) is the standard frequency measurement and means "cycles per second." So, a sampling frequency of 30 Hz means 30 samples per second, or once every 30.3 ms, whereas a sampling frequency of 2,000 Hz means a data point is collected every half millisecond (500 microseconds). Let's use a sampling rate of 300 Hz (i.e., every 3.3 ms) as our example.

One algorithm to identify saccades is the Velocity-Threshold Identification (I-VT). Velocity is calculated in degrees per second (i.e., the time it takes for the eye to move by a given angle, aka, angular velocity). Here too a threshold is chosen above, which an eye movement is considered to be a saccade. For example, in one of the eye trackers we use in our lab, which are made by Tobii, the threshold is 30 degrees per second (Olsen, 2012, p. 8).

Once we have identified the saccades, the periods between the saccades are considered fixations. Any series of fixations that are uninterrupted by saccades are added up to calculate the length of the fixation. So, if, for example, at time t_0 we have the first fixation, and we have a string of uninterrupted data points under the velocity threshold until time t_7 our fixation length would be 26.4 ms, assuming a sampling rate of 300 Hz.[1] In reality the process is much more complicated, and various tests are performed to make sure that blinks, for example, are not considered saccades, or that the points in the fixations are within a certain threshold of each other (between 0.033 and 2 degrees) (Komoortsev et al., 2010, p. 2640). This latter method, based on a region, is also used to identify fixations on its own (these are called dispersion algorithms). Salvucci and Goldberg (2000) and Duchowski (2017, pp. 141–74) provide more in-depth reviews of various fixation identification algorithms.[2]

Compensatory Movements

We have discussed two types of movement so far: saccades and fixations. However, they are not the only types of movements that the eyes are capable of. In this section present sections we will review compensatory movements and in the following one the small movements that occur *during* a fixation.

Compensatory movements are movements of the eyes performed to compensate for the movements of the head or the body. If you turn your body or just your head, but keep looking at something, for example, the person you are talking to, the visual system compensates for that movement in order to keep your target focused. Less dramatically, head movements, such as tilting the head left or right, are routinely compensated for by torsional movements. The general purpose of compensatory movements is to stabilize the image, for example, while walking. If you have ever watched a video shot while walking without a stabilizer, you get a good feel of what the raw input of the eyes would look like without compensatory movements.

Vergence

If we are looking at an object sufficiently far away, the gaze of each eye can be considered parallel and so the eyes move in conjunction. The reader will recall Hering's law of equal innervation (Chapter 1) and the principle of binocular vision. However, when we are looking at objects near the eye, the difference in angles between the eyes may be significant. **Vergence** can be defined as the eyes moving in different directions to accommodate objects closer to the eye or if one object is nearer to the eye. For example, if you fixate on an object close to your eyes and in the center of the optical field, such as the tip of your nose, the eyes move separately in opposite directions (the right eye moves to the left and the left eye moves to the right). Vergence movements are much slower than saccades. Symmetrical vergence, like the example discussed, is relatively uncommon, and generally saccades and head movements will accompany vergence to accommodate the position of the object. This happens when we move from looking at an object far away to one nearby, for example.

Pursuit Movement

So far, we have more or less implicitly assumed that the target of the fixation is static. **Pursuit** movements occur when the eye follows a moving target. They are usually fairly slow and can range between 1°and 30°. Pursuit movements consist of smooth slow tracking and saccades. The smooth tracking movements try to match the motion of the moving target, whereas the saccades recenter the fovea on the target if the target is no longer in the fovea. Pursuit movements are partly voluntary (e.g., if one is watching a tennis match, one can follow the ball deliberately, among a variety of other moving objects, such as the players, the spectators, etc.). However, there is an automatic aspect to pursuit, because retinal motion detection can trigger eye movements autonomously. This makes evolutionary sense, as automatically spotting moving predators carries an obvious evolutionary advantage over having to decide to do so. A latency of 70–130 ms has been reported for pursuit movements to start, after a stimulus starts moving.

Pursuit movements also include an anticipatory or predictive component. For example, if a stimulus disappears, as in the case of a rabbit running behind a tree, we "remember" its speed and we continue to follow the path we predict it would have taken. We can also do predictive pursuit of targets that move regularly, such as a pendulum (e.g., a child in a swing). In conclusion, it is clear the pursuit movements involve both retinal information (e.g., the velocity of the movement, the rate of acceleration) and extra-retinal information (i.e., input from areas of the brain), including expectations, attention, and memory (i.e., prior behavior of the target). Indeed, it is very difficult, if not impossible, to activate **smooth pursuit** ocular movements in the absence of a moving target.

Miniature (Fixational) Eye Movements

Under the label "miniature eye movements" are included those movements that occur *within* a fixation (as such they are also known as "fixational eye movements"), that is, microsaccades, drifts, and tremors. Miniature eye movements are very small and generally remain within the foveal area. All these movements are unconscious and are not controllable volitionally. The best source on miniature eye movements is Martinez-Conde and Macknik (2007), which is very accessible. Martinez-Conde and Macknik (2017) is more advanced.

Microsaccades

Microsaccades (also called "flicks") are usually defined as involving movements of the eye of less than 1°. The boundary between microsaccades and saccades is not absolute, meaning some microsaccades are larger than 1°, but the vast majority (over 90 percent to be accurate) is captured by that threshold (Martinez-Conde & Macknik, 2011, p. 100). Microsaccades are actually quite similar to saccades: they are generally binocular (involve both eyes), they involve visual suppression, and they are planned in the same areas of the brain.

As we saw in Chapter 3, microsaccades have the function of keeping the orientation of the fovea moving, because after a period of stillness vision fades, since motion is one of the factors that causes vision cells to fire. The Troxler effect (see Chapter 1) is closely related to this phenomenon. Microsaccades are usually 30 ms apart.

Microsaccades are not controlled consciously (unlike saccades which may be controlled), and they have been linked to shifts in covert attention (see Eye Tracking and Attention section). Essentially, it is as if the brain is planning to look at a given target and starts "stealing" microsaccades in that direction. They may be used to refocus and in fact they may constitute a way of sampling the visual field. Recent work has connected microsaccades and attention, showing that high visual attention correlates with high microsaccade rate (e.g., Krueger et al., 2019). Finally, we should mention that microsaccades have been connected to "enhanced visibility of peripheral and parafoveal stimuli" (Martinez-Conde & Macknik, 2011; McCamy et al., 2012). This will be quite significant when we tackle reading (in Chapter 5).

Drift

Drift is in a sense the opposite of a microsaccade, as it represents the tendency of the fovea to slowly move away from the locus of the fixation. Drift is random and very slow and remains within the fovea (Ko et al., 2016, p. 2; Yarbus, 1967, p. 108). Note the randomness of the drift in Figure 4.4. It is unclear to what extent drift is binocular (Otero-Millan et al., 2014).

Tremors

Tremors are very fast oscillations superimposed on drifts. They can be visualized as a zig-zag motion that occurs during the drift. To use a metaphor, a drunken person attempting to walk on a straight line may oscillate to the

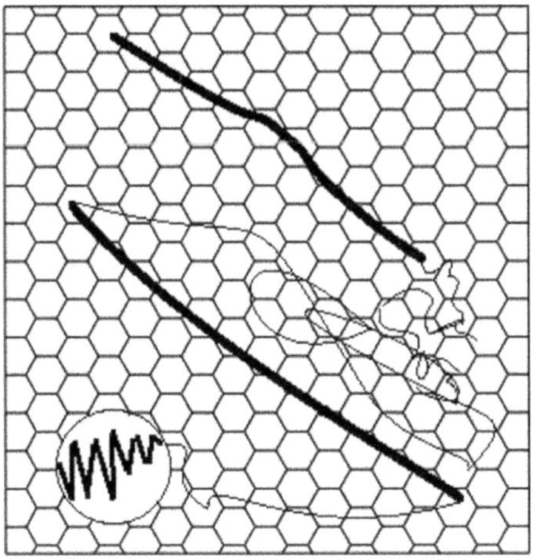

FIGURE 4.4 *Microsaccades (thicker line) and drift (thinner line). The hexagons represent the retinal photoreceptor cells. The magnified detail in the circle represents the tremor.*

left and to the right of the line (tremor) and wander off away from the line (drift) while still oscillating on either side of the drift. A sober person will just walk on the straight line (microsaccade). Tremors are the smallest movements in the eyes. Their width is about the size of one photoreceptor cell. Tremors fall below the threshold of most eye trackers, because of their small size and fast frequency (i.e., how fast they oscillate on either side of the drift). There may be some degree of binocularity, but it appears to be small (Otero-Millan et al., 2014).

Eye Tracking and Attention

We saw in the introduction the fundamental assumption that gaze equals attention in the psychological applications[3] of eye tracking (and linguistics counts as a psychological science in this context). Researchers distinguish between **overt attention** and **covert attention**. Intuitively, it is fairly easy to grasp the difference: anyone who has sat in a meeting listening to an uninteresting topic will have experienced how easily, while directing one's overt attention to the Provost's ideas on faculty duties (for example), one can be in fact thinking about their shopping list (covert attention). The phenomenon of mind wandering (see Chapter 8) is of course not the only instance of decoupling of overt and covert attention. Coming to eye movements and attention, there is evidence that when we are (subconsciously) planning the next saccade, covert attention goes to the planning of the saccade. Godfroid (2020, pp. 19–21) reviewed some of the literature on the subject.

An important aspect of the interplay of attention and visual perception is revealed by the phenomenon of **inattentional blindness**. Inattentional blindness has been researched in numerous experiments but the best known is the "the invisible gorilla" experiment (Simons & Chabris, 1999; Chabris & Simons, 2010; Figure 4.5). In the experiment, subjects were shown a video of a group of people playing basketball and were instructed to count the number of passes of one or the other team. At some point, a person in a gorilla suit walked by, stopped in the middle of the frame, briefly thumped its chest, and walked away. Amazingly, about 50 percent of the subjects did not notice the gorilla. It should be noted that if a subject counted the passes wrong, they were rejected from the experiment; in other words, only those who had been *really* focused on the task were considered.

Nonetheless, the experiment, and many others, showed convincingly that attention is finite and that if a significant load is put on it, even a very salient and completely visible stimulus fails to be seen. Further studies showed that many participants failed to see a target item, despite having actually fixated on the object. In other words, as has often been noted, there is a significant difference between looking at something and seeing it. The difference is

FIGURE 4.5 *Frame captures from the invisible gorilla experiment.* Source: *Simons and Chabris (1999). Figure provided by Daniel Simons: www.dansimons.com.*

visual attention. The field of visual attention is too complex to review here, so we will refer briefly only to those theories that are widely used in eye tracking, within linguistics, which are the spotlight theory, the zoom lens theory, and especially the eye-mind theory.

The Spotlight Theory

The basic idea of the spotlight theory is simple: much like if one is observing a large object, say the facade of a house, at night, with only a flashlight, one has to make a choice of what to observe: a decision has to be made of whether to look at the door or at the roof, for example. So attention is directed to a subset of the visual field. However, Posner et al. (1980) separated attention and foveal focus, as we saw by differentiating between covert and overt attention. The size of the spotlight may change: for example, one experiment found that subjects could focus on one letter at a time (1°) and up to five letters. Downing and Pinker (1985) found that the spotlight is focused gradually more on the foveal region and less in the peripheral region (and so they speak of a "peephole"). This sort of evidence has led to the formulation of a variant of the spotlight theory, the **zoom lens theory** (Eriksen & St. James, 1986), which claims that the focus of attention is variable, precisely like a zoom lens.

Bottom-Up and Top-Down Attention

The **early selection** approach holds that perception is inherently limited and that it requires attention to proceed. Therefore what to focus on is selected early on in the process on the basis of a rapid analysis of perceptual features (bottom-up). Saliency (or salience, i.e., the property of being salient) is a central part of this process. For example, a red dot in a field of black dots is salient (stands out) and thus will be selected for attention. Stimuli that fail to make the cut are either ignored (recall the invisible gorilla) or not fully perceived. The selection process is often referred to as the "bottleneck" of processing.

The **late selection** approach, conversely, assumes that perception is essentially unlimited and thus selection can be done at a later stage, after full perception. Selection of the stimulus to attend to will thus be influenced by higher cognitive processes, such as memory (recognition) or task (recall Yarbus's finding that the distribution of the points of fixation varies according to the task provided to the observer). Late selection is thus a top-down approach. Significant debate between the early and late selection approaches has endured (see Driver, 2001, for a review).

The Eye-Mind Theory

Just and Carpenter (1980) proposed a theory of reading (based on scientific reading, but there seems to be no reason not to generalize), which assumes that "the eye remains fixated on a word as long as the word is being processed. So the time it takes to process a newly fixated word is directly indicated by the gaze duration" (p. 330). So, one can summarize the eye-mind theory as: what is being read by the eye is processed by the mind. A further assumption of the eye-mind theory is that processing is immediate: there is no delay between fixation and processing. It is fair to say that the eye-mind assumption for reading is what the gaze equals attention slogan is for eye tracking at large.

However, the eye-mind theory in an absolute formulation is clearly falsified by a host of phenomena in which the position of the foveal vision does not match what is currently being processed: for example, spillover effects (Pupil Dilation section), parafoveal preview (Chapter 5), mind wandering (Chapter 8), and covert attention (See Eye Tracking and Attention section). So a better formulation is that the mind tends to process what the eyes are gazing at, although there may be temporary decouplings between gaze and attention.

Conclusion: Vision and Attention

This section had the fundamental goal of problematizing the "gaze equals attention" informal motto of eye tracking. First, while gaze does correlate with attention, other factors come into play. One is the nature of the stimuli. Another is the cognitive load of the perceiver. Second, attention is not a unitary phenomenon; it can be analyzed into overt and covert attention. Third, processing of visual information is not itself a transparent process; it may have a significant impact on what (part of the) data gets processed at all and to what level it gets processed. Those wanting more information on the theory of attention should consult some of the books written specifically on the subject, such as Styles (2006). See also Chapter 9 on visual processing, which has more discussion of some of these topics.

Eye-Tracking Metrics

We are now ready to approach in more detail the nitty-gritty details of eye tracking. In particular, the following sections will consider various metrics that are routinely collected and analyzed by eye trackers. The purpose of this section is to familiarize the reader with the variety of measurements that are available and some of their uses. This is not meant as a comprehensive treatment. In particular we should stress that each eye tracker has its own set of metrics, and so it is recommended that researchers should familiarize themselves with the metrics offered by their specific equipment.

Areas/Regions of Interest

Areas or regions of interest (**AOIs/ROIs**) are user-defined regions of the visual field that are of particular relevance to the researcher. Their significance lies in the fact that several metrics can be defined relative to an AOI/ROI. For example, Number of Fixations and Fixation Time can be relativized to an AOI/ROI; in other words, we can get a metric of how often the user fixated within an AOI/ROI or how long they spent within it. The specifics of how to set up AOIs/ROIs will obviously vary among eye-tracking system to system, but the general principle is the same. AOIs/ROIs are determined by the goals of the study and by the nature of the visual field under observation. Suppose, for example, that you are interested, as Simarro et al. (2020) were, in the order of processing of tweets. We divided the images into four AOIs: the image, the caption, the user name, and a catch-all category of "outside" (meaning anything else). Other studies may suggest other arrangements: to use another example, Gironzetti (2017) when examining the effects of gaze and smiling in face-to-face encounters used AOIs for the mouth and the eyes.[4]

One thing to keep in mind is that one does not want too broad of an AOI (although some systems let you define AOIs/ROIs after the fact and so one could always change an overlarge AOI/ROI[5]), which may capture fixations that are not really on target. Conversely, since foveal vision has a range of about 1.5°, which corresponds to about 1 centimeter (about one-fourth of an inch), if looking at a computer monitor at 60 cm of distance (roughly

24 inches), two adjacent AOIs/ROIs of smaller size would be impossible to discriminate, lacking an extremely precise eye tracker. Relatedly there should be a space of 1° between AOIs/ROIs, if precise discrimination is needed. Holmqvist and Andersson (2017, p. 262) recommend a minimum size of 1° to 1.5° for high-end trackers, which means an even larger size for less powerful trackers. In conclusion, you should be very careful when designing your AOI/ROI and be cognizant of the limits of precision of the eye-tracking system you are using. Here too, as we will discuss in the methodology section (Chapter 10), it is crucial that you be familiar with the standards used in your subfield and you should adhere to them. Using an eye tracker with insufficient sampling rate for your data may invalidate your study completely. As a preliminary phase before setting your AOIs/ROIs you may want to run a cluster visualization of your data, which can help you see where the data tend to group and may be useful in preparing the AOIs/ROIs. Some recent models of eye trackers even have dynamic AOIs/ROIs, which will "follow" an AOI/ROI in a video, for example. Holmqvist and Andersson (2017, pp. 254–301) have a very detailed and useful section on setting up AOIs/ROIs and how to handle difficult situations (e.g., overlapping AOIs/ROIs).

AOI/ROI metrics vary, depending on the purposes of the study. The simplest is a fixation count (i.e., the number of times a fixation falls within the AOI; see later in the text), but other metrics may include skips (when an AOI/ROI is not fixated) and returns (when a participant leaves an AOI/ROI and then returns to it).

Fixations

There are numerous fixations-related metrics and more can be developed for study-specific purposes. The following reviews some of the most common ones. It is a good investment of time to thoroughly familiarize yourself with the type of fixation metrics your equipment provides by default.

Fixation/Dwell Time

Generally speaking, **dwell time** and **fixation time/duration** are very closely related. Dwell time is defined as the amount of time that a user spends in a given location or within an AOI/ROI, whereas fixation time is dwell time, minus the time spent in saccades. Saccades take little time, so, as we said, the two are usually close. Dwell time is often used in reading experimental work, since it has been shown that higher dwell time signals increased processing load (see Chapter 5). Generally, it is possible to differentiate between each individual visit duration and an overall duration (i.e., the sum of all visits). This can be useful, for example, if the subject returns to the same area repeatedly. **First fixation duration** is a subset of fixation time, which applies only to first fixations, so it measures only the duration of the first fixation in an AOI. This can be useful, for example, in reading, to identify errors (which are associated with longer first fixation durations).

Time to First Fixation

This metric is fairly clear intuitively as it measures the time elapsed before the subject fixates on an AOI/ROI. In the Simarro et al. (2020) study we mentioned earlier, we wanted to know in what order the participants accessed the AOIs, but we also wanted to know how long it took them to do so, and so we made extensive use of this metric. Other uses may involve how long it takes for a user to spot a button they need to click on in a website or how fast they spot a given feature of interest (e.g., a spelling mistake).

Mean Fixation Duration

Mean fixation duration represents, as the name indicates, the average length of fixation, either for an individual subject or averaged over all the participants in an experiment. It is usually taken to correlate with cognitive load (in other words, we look longer at harder-to-process targets).

Fixation Density/Coverage

This metric examines how much of the visual field has been covered by fixations. Fixation density (also known as fixation dispersion) essentially correlates the number of fixations into an AOI with its size. This is of particular use in human-computer interface studies and in web page usability studies.

Repeat Fixations

Repeat fixations or refixations take place when an element is fixated more than once. Refixations can be indicative of the significance of a given element.

Regressions

A **regression** fixation is when a subject fixates on point *a*, moves on to point *b*, and then returns toward or to point *a*. Regressions are of particular significance in reading, as they often denote processing problems (e.g., garden-path sentences) or just repairs for overshooting the target in a saccade (see Chapter 5).

Gaze Paths

Gaze paths (aka, scan paths) are an ordered series of fixations (circles) and saccades (lines). They record the process of scanning a scene or the progress along a text. They are particularly useful to reveal the order of access of

FIGURE 4.6 *Scan or gaze path.*

various components of a scene and/or backtracking during reading. The size of the circles is proportional to the duration of the fixation.

The idea of a scan path was introduced by Yarbus (1967) and Noton and Stark (1971). Yarbus (1967) produced numerous plates, which reproduced all the saccades and fixations of one or several subject(s) observing the image over a period of a few minutes. While Yarbus does not use the term "scan path" (he speaks of the "distribution of the points of fixation"), clearly those images are the precursor of today's gaze paths. Noton and Stark's definition was much more complex than the one presented earlier, as they asserted that the scan path was an important component of the processing of a visual pattern and that that sequence was crucial to the recall of the pattern. These claims have been taken up in education theory (see Chapter 8).

Figure 4.6 shows a gaze path, generated by one participant in Simarro et al. (2020) looking at the image. The gaze path covers the first ten fixations and the saccades between them, for a total duration of 3.8 seconds. Clearly the participant first read the caption, then looked at the sign, and then looked at the cat.

Gaze plots may represent one or more users, usually differentiated by color; however, one of the drawbacks of gaze plots is that they become crowded and confusing when they try to present too much data at the same time, especially if fixations overlap. Here minimalism (i.e., using the least amount of information to make a point) should be the goal.

Heat Maps

Heat maps are visualizations that use a color scale to graphically represent the areas of the visual field that attract the most attention (see Figure 4.7).

The quantity of attention mediated via gaze can be quantified differently, using number of fixations or duration of fixations, or relative duration of fixations. The relative duration is not as intuitively clear, but it shows the percentage of the time spent by the subjects on the area. Suppose that you have subjects looking at tweets, as in our case, and they can freely decide how much time they will spend on each image. Suppose further that participant A spent 50 percent of a 4-second visit looking at an AOI, whereas participant B spent 5 percent of a 40-second visit. Clearly A was much more interested in the AOI than B was, but the absolute time spent would be the same for both of the hypothetical participants (2 seconds). The relative duration gives you a tool to differentiate between these two scenarios.

Unlike gaze paths, heat maps are summative: they display the overall amount of time or amount of fixations spent by a participant or several, onto the visual field. Thus no order of fixation is available in a heat map. By their very nature, they are not very precise, as it can be difficult to assess the specific value associated with a given color at a given point. However, recent online interactive heat maps allow you to click on a point and get a value.

Note the mapping legend in the bottom of the image: transparency (no color) is indicated as is common in high-end image processing software by a gray squares grid. Note also the intuitive color palette where red represents the highest values and orange, yellow, and green intermediate values. No color indicates the lowest values. Heat maps can be generated in grayscale or as opacity maps, in which the opposite convention applies: the highest values are transparent (and thus can be seen), and the lowest values are darkened out (see Figure 4.8).

FIGURE 4.7 *Absolute duration of fixations heat map.*

FIGURE 4.8 *An opacity map of the same tweet as in Figure 4.7.*

Heat maps are very popular, due to the fact that they are easy to understand and do not require any sophisticated reasoning: warmer colors mean more fixations. However, possibly because of this, there is a distrust of heat maps for scientific work, unless it is very clear what one is doing. While most users will be content with the heat maps generated by their eye-tracking software, it is possible using more advanced data visualization software to produce your own heat maps, using customized color mappings. Even without resorting to such advanced tools, a simple heat map as in Figure 4.7 may be appropriate if sufficient care is given in providing the information to disambiguate the meaning of the heat map. For example, it obviously makes a huge difference if we are looking at one participant or at all participants.

Heat maps, like any other kind of visual display (e.g., charts and graphs), should be used to make a clear point and to illustrate factual information: for example, in Figure 4.7, it is immediately apparent that the duration of fixations of this subject matches the significant parts of the humor mechanism: the word "legal loophole" (in Spanish), the no dogs allowed sign on the grass, and of course the cat.[6]

Saccadic Measurements

Saccade rate is a metric of the frequency of saccades. It is used in search tasks, where it has been observed that a faster and less accurate saccadic rate may be preferred over a slower more deliberate, and more precise, saccadic rate (e.g., Wu & Kowler, 2013). Microsaccade rates are also measured, but of course they require very sensitive eye-tracking equipment. Microsaccade rates are typically in the range of 1–3 microsaccades per second. Other measurements include the length and the speed of the saccades.

Pupil Dilation

Pupillometry, that is, the measurement of the size of the pupil is a separate field of inquiry, but it is often pursued within eye tracking because eye trackers collect information on pupil size and it is easily retrievable.[7] The interest in measuring the size of the pupil lies in the theory that pupil dilation matches cognitive and emotional arousal. However, changes in pupil dilation are not limited to arousal: for example, a fairly well-established result is that the pupil dilation correlates with increased difficulty of a task and/or mental effort (see a review of the evidence, in van Steenbergen, 2018). An interesting example is Laeng et al. (2011), who showed that the cognitive load of the Stroop effect (see Chapter 6) is reflected in pupillary dilation. Recent work on pupillometry has also shown a correlation with decision making (Strauch et al., 2018). One of the biggest problems with pupillometry is that the pupil is very responsive to light (see Chapter 2), and so great care should be taken that all visuals have the same brightness. Furthermore, changes related to brightness are much stronger than changes related to emotional or cognitive load. One of the advantages of using pupillary dilation data is that pupil dilation is not subject to conscious control and thus cannot be faked. A good recent survey of pupillometry is provided by Laeng and Alnaes (2019).

Other Non-Eye-Related Metrics

Increasingly researchers are associating eye tracking with other physiological metrics in an effort to gain an even better understanding of what the brain is doing while perceiving a scene. These metrics are outside the scope of this book, as they require an entirely different set of knowledge and skills to administer, but we will mention the following:

1. **Galvanic skin response** (GSR) falls under the broader category of **electrodermal activity**, that is, electric current or potential on the skin. We already saw the use of electric potential around the eyes as a way of tracking gaze (Chapter 1). Galvanic skin response measures very small electric current on the skin, usually in the palm of the hand. Galvanic skin response is very closely related to the activity of sweat glands and hence it provides a direct way to assess one's emotional arousal (e.g., stress) and cognitive load. It

is used, in combination with eye tracking, in marketing, human-computer interaction, but also more generally where assessment of a subject's emotional status is important, such as in the diagnosis of depression. Depressed patients have lower galvanic skin response and tend to look longer at negative images (Ding et al., 2019). It is sometimes used in correlation with pupillometry.

2. **Electroencephalography** (EEG) is the measurement of the electrical activity of the brain, usually done with electrodes attached to the scalp. In some cases electrodes are inserted into the brain, for example, during brain surgery. Its diagnostic uses include diagnosing epilepsy, brain tumors, and sleep disorders. EEG has low spatial resolution (since it records through the skull) but it has excellent temporal resolution (in the milliseconds), so it is not very precise as to the location of the electrical activity, but it is very precise as to the timing of it. EEG has a sampling rate between 250 and 200 Hz, which is comparable to eye tracking. In connection with eye tracking, it is used mostly in the **event-related potential** (ERP) paradigm, that is, a change in EEG pattern as a response to a stimulus. Particularly relevant to eye tracking are ERPs related to saccades, microsaccades, and fixations. For example, Meyberg et al. (2015) recorded microsaccade-related potentials (mSRP) in a cueing experiment and found that microsaccades were biased toward the cues direction and that the amplitude of the mSRP was greater. This is significant, because we know that attention to a stimulus increases ERP and this suggests that microsaccades covertly orienting toward a stimulus do so as well. Microsaccades could thus be cues to covert attention shift. On EEG and eye tracking see also Holmqvist and Andersson (2017, pp. 363–4, 372).
3. fMRI: In the past twenty years, functional magnetic resonance imaging (fMRI) has been jointly performed accompanying eye tracking, and there are now fMRI machines that incorporate an eye-tracking component as well. See Duchowski (2017, pp. 253–4) for a review of some studies.

Conclusion

We have reviewed the principal constructs used in eye tracking, beginning with eye movements: saccades, fixations, compensatory movements, smooth pursuit movements, and minor movements that occur within a fixation (microsaccades, drift, tremor). We then reviewed some of the approaches to attention, in particular distinguishing between overt and covert attention. Finally, we covered a variety of metrics used in eye tracking. Taken together, these concepts form the basis of eye-tracking research. This concludes the first part of the book as well. We have presented the historical, physiological, neurological, and theoretical foundations of eye tracking. We can now turn to Part II of the book, where we will discuss some of the applications of eye tracking and their findings.

Self-Study Questions

The following questions pertain to the contents of this chapter. You should be able to answer them without looking up the answer in the book.

1. What is the difference between saccades and microsaccades?
2. What does it mean that fixations are an abstraction?
3. What does the invisible gorilla experiment prove?
4. What is the difference between overt and covert attention? How can covert attention be inferred?
5. What is the advantage of combining eye tracking with other physiological metrics?
6. Explain some of the concerns over the use of heat maps.
7. What is the difference between bottom-up and top-down attention?

Words to Know

saccade	fixation	vergence	smooth pursuit	microsaccade
saccadic suppression	drift	tremors	scan path theory	pupillometry
AOI	ROI	regression	gaze path	heat map

Further Reading and Sources

The first section of the chapter is largely based on the relevant chapters in Liversedge et al. (2011), which is an outstanding resource but rather advanced. The discussion of attention is based largely on Styles (2006), Driver (2001), and on the references quoted in the text. Janisse (1977) is an excellent introduction to pupillometry. The discussion of eye-tracking methods is based on Holmqvist and Andersson (2017) and Duchowski (2017).

Part II

Applications

CHAPTER 5

Reading

Reading is a very important activity as in literate societies it is arguably the main channel of cultural transmission. However, in this context we are not concerned with the cultural or social significance of reading but rather with its mechanics. Therefore it will be useful to start by presenting a very short and simplified overview of the process. For ease of discussion, we will discuss the reading of English, with the caveat that there are differences for other languages.[1]

Hierarchical Model of Reading

Reading is a very common activity, at least in literate societies. We seldom think about it after we master the practice usually by the end of elementary school.[2] A model of reading incorporates the following stages/processes:

1. Perceptual level: raw perception of letters
2. Lexical level: meaning of words is extracted
3. Propositional level: clauses and propositions formed
4. Situational model: representation of the mental model formed by the hearer

Obviously, one does not stop after the text has been represented in the mind, but the processes of interpretation, critical assessment, and so on are no longer specific to reading (they are, for example, shared with interpretive dance).

Let us briefly take up each step of the process, to set the background of the discussion: first, one must perceive the letters, in their physicality (i.e., the letter "o" is a circle, the letter "p" is a circle with a tangent line, etc.).

Balota et al. (2006) provide a clear synthesis of the research. Essentially, there are neural substrates that specialize in recognizing vertical and horizontal lines, angles, and so on. These features are integrated in letters (so a "t" would be integrated into a pattern consisting of a vertical line with a quarter of a circle at the bottom and a short horizontal line crossing the vertical line). Letters are then integrated into words. Interestingly, it is easier to identify letters within a word than in isolation (Balota et al., 2006). While this is an interesting area, eye tracking has not been central to it.

The step from letters to words includes the notoriously thorny issue of the phonotactics of English whereby, for example, the letter "c" corresponds to the phoneme /k/[3] but to the phoneme /s/ when followed by "i, e," and "y" or the sequence "mb" is read [mb] in the middle of a word but [m] at the end of one. We refer to languages which are close to a one-to-one correspondence between graphemes and phonemes (such as Swahili or Serbo-Croatian) as shallow orthographies and languages (such as French and English) which have complex grapheme/phoneme connections as deep orthographies. Another term, probably clearer, is transparent and opaque systems. The significance in the distinction comes from the fact that the research points to two different routes from spelling to word: in transparent/shallow languages the speakers go from spelling to word via the phonetic rules we already discussed. In opaque/deep languages there exists a second route that consists of accessing the word as a whole directly. There is also evidence that the onset of a syllable (any consonant to the left of the vowel) and the rime (the vowel and following consonants, if any) play a role on word recognition (sequences of letters that do occur in words are recognized faster than those that do not).

Syllables and morphemes also play a role in word recognition. For example, priming experiments have shown that "dishonest" primes for "honest" but "arson" does not prime for "son" because "son" is not a morpheme (unlike in "stepson," for example, where "son" is a morpheme). At the word level, research has shown that length plays a role (longer words are harder to read), as do frequency (more frequent words are easier to recognize), familiarity (familiar words are easier), age of acquisition (words acquired early are easier), and similarity to other words (words spelled like many others are harder, generally speaking, but this also goes for words that sound like others). Once words have been identified they are assembled in sentences and their meaning is evaluated (typically represented as propositions).

Finally, the sentences and their context are integrated in a mental representation of what the sentences have conveyed to the hearer. This is called a mental or situation model. Mental models are similar to frames or scripts (see Attardo, 2020 for a review), but they are specific to the utterances and their context, whereas frames are more abstract. Zwaan and Radvansky (1998) sum up this stage of the reading process: "the construction of a coherent situation model is tantamount to the successful comprehension of a text" (p. 163).

Eye Tracking and Reading

Eye tracking has made a significant contribution to the study of reading. However, if we look at the previously discussed hierarchical model, we will see that eye tracking has contributed more at the lexical-propositional level, as one would expect. So, let us start with a quick vignette of what actually happens when we are reading.

During reading, the reader moves their foveal vision area from left to right on the line of text. The movement is not smooth and continuous; it is done by saccades and fixations. Fixations are indicated in Figure 5.1 as red circles, numbered in the order of fixation. The saccades are the lines connecting them. The typical fixation covers one content word, but for longer words we can see two fixations. So a typical line of text will require four or five fixations/saccades. Generally small, high-frequency function words, such as pronouns and prepositions, are not fixated (they are fixated 35 percent of the time, as opposed to 85 percent for content words; Rayner, 1998, p. 375). See, for example, the skipped function words in Figure 5.1: the three instances of "the," "in," "that," and "is." Not all saccades are in the left to right direction; about 15 percent are in the opposite direction. **Regression** saccades are returns to a previously fixated word. An example is found in Figure 5.1: the regression takes place from fixation 4 (in the center of "surrounding") to fixation 5, at the beginning of the word. Regressions may be corrective (and in that case typically very short) if for example the saccade overshot the **optimal viewing position**, that is, the center of the word or slightly to the right of the center. Alternatively, they may reveal a problem in the comprehension or processing of the sentence. In the latter case, they are longer. Another kind of right-to-left saccade takes the reader to the next line. When the reader has reached the end of the line a **return sweep** (i.e., a long saccade to the beginning of the line below) is performed (see an example in Figure 5.1, between fixations 6 and 7). This is often followed by a corrective saccade (the reader will recall that saccades are ballistic) because return sweeps often undershoot their target. There is variation which is tied to proficiency: children initiate the return sweeps closer to the end of the line than adults do and require more corrective saccades (Parker et al., 2019).

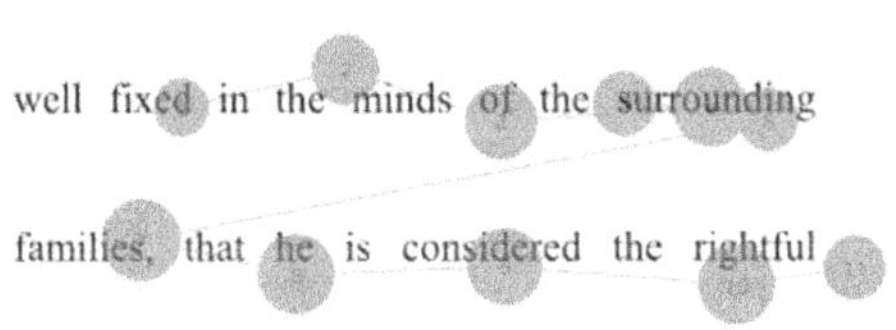

FIGURE 5.1 *A sample scan path of part of a sentence, from Jane Austen's* Pride and Prejudice.

As the reader will recall, during saccades we are practically blind; hence, the acquisition of the information from the words on the page must take place during fixations. Obviously, the foveal region, which is the one in greater focus, is the main source of information, but the parafoveal region does acquire some information as well, which is used for the planning of the next saccade, for example. We will consider both foveal and parafoveal vision in what follows. According to the eye-mind hypothesis (see Chapter 4) there is a correlation between attention and foveal vision location. Consequently, longer fixations are taken as evidence of the need for longer processing. The need for longer processing may be caused by various factors which we will consider in what follows. We will first focus on saccades and fixations while reading; we will then consider the targeting movements for saccades and the various factors that affect processing time during fixations. Finally we will wrap up the discussion by briefly considering some of the models that have been proposed to integrate the process.

Perceptual-Level Phenomena

Since the stimuli of reading are typically fairly small (a twelve-point letter is one-sixth of an inch tall, i.e., 4.2 mm), reading research requires powerful eye trackers with a resolution of at least 1,000 Hz. Also, both saccades and fixations tend to be shorter than in visual processing, so very high precision and accuracy are required, or the reader's fixations may be misattributed.

Reading is sensitive to factors such as font size and type. Beymer, Russell, and Orton (2008) concluded that the smaller the font, the slower subjects read. Legge et al. (1985) showed that there is an optimal window of readability, between 0.3° and 2°, or in other words that characters that are too small or too big are hard to read (recall that the angular width of the letters is dependent on the distance from the eye, see Chapter 4). Wittily, Legge et al. provided their readers with instructions:

> At a typical reading distance of 40 cm, [15 and 3/4 in], the character size is 3° in the title letters on the cover of this journal, 0.4° in the title of this article, and 0.2° in this text. If you are in a hurry, read this article at 27 cm [10 1/2 in] or closer to increase the character size to at least 0.3°, for maximum reading rate. (1985, p. 247)

Legge and Bigelow (2011) provided a very useful discussion and helpful charts to convert point sizes to angles. They gave as an operational range for maximum reading speed of x-height (i.e., the height of the letter "x" in a font) between 4 point size to 40 points (1.4 mm to 14 mm) (2011, p. 5). Newspapers use anywhere between 4 and 5 point fonts for regular text. Rayner et al. (2010) reported that the fonts do not make a difference (proportional versus fixed width[4]), but that reducing the spacing between the letters actually improves reading speed.

Visual Acuity

One may wonder why reading has to be such a complex process, with several fixations in one line. Why not do one sweeping fixation per line of text? It turns out that is just impossible, because of visual acuity.

Visual Acuity

As the reader will recall, visual acuity, that is, the capacity to tell letters apart, in reading, is highest in the foveal region. The foveal region corresponds to about 2°, which roughly corresponds to six to eight letters. The parafovea is about 5°. Parafoveal vision has also some impact on reading, but it is very difficult to read with parafoveal information alone and it is impossible to do so with peripheral information only (Rayner, 1998, p. 375). In other words, the further away from the center of the foveal region the letters are located, the harder it becomes to read them. Once we get into the peripheral region, all that the reader can identify is the general shape of the word (e.g., is it long or short?). Because of this, it becomes necessary to move the foveal region along the line of text, to be able to read the letters.

Visual and Perceptual Span

The **visual span** is defined as the number of characters that are identifiable at a glance, in reading. Legge et al. (1997) estimated that the normal visual span has a range of between ten and two characters, depending on a series of factors, including contrast and size. Individual variation tied to quality of vision was also present.

The perceptual span is wider and includes parafoveal vision. According to Rayner and Pollatsek (2006) the perceptual span extends fourteen to fifteen characters to the right and four to the left of the fixation. The perceptual span extends only to the line being read and is consistent across alphabetic languages (in other words, this does not apply to Chinese, for example). In Arabic and Hebrew, which are alphabetic, but read right to left, we find the same asymmetry, but in their direction of reading: so in Arabic the perceptual span is larger to the *left* (Hyönä & Kakkinen 2019). Orthography, word length, and complexity affect the perceptual span. There is also individual variation, as better readers have longer perceptual spans. Attention also affects perceptual span length. Developmentally, this is very clear as, for example, second graders have a smaller (eleven characters) span, which grows to fourteen characters by sixth grade.

Saccades

As we saw, the reader does not sweep gently along the line of text but advances by saccades. Saccades in reading are very short: typically they last 20–40 ms. Their duration correlates with their length. Saccades are typically between seven and nine characters; shorter saccades tend to be corrective and may be as short as one character; longer saccades of up to twenty characters are also encountered, but they tend to occur after a regression, when the reader returns to reading at a point *ahead* of the point from where the regression was launched (Rayner & Pollatsek, 2006). Return sweeps, as one would expect, are the longest saccades in reading. The difficulty of the text also affects the length of the saccades: complex texts cause shorter saccades and longer fixations. Generally, saccades move on to the following word, but for very long/complex words, two fixations may take place on the same word (words may also be skipped, in longer saccades, and we will return to this in the Skipping section). Thus there is great variability in the length of saccades in reading, as the range is one to twenty character spaces.

Fixations

There is extreme variability in the duration of fixation in reading. According to Rayner and Pollatsek (2006) a typical fixation lasts 200–250 ms, but fixations may be as short as 50 ms and as long as 500 ms (p. 614). The location of the first fixation is usually five to seven characters from the beginning of the line, and the last fixation is usually also five to seven characters from the end of the line. Fixation length is affected by a variety of factors, which will be reviewed in Lexical Influences section.

Saccade Latency

According to Reichle et al. (2003, p. 446), "the time needed to plan and execute a saccade, is approximately 180–250 msec." However, since an average reading rate is 300 words/minute, "the decision to make a saccade is often made within the first 100 msec of a fixation" (p. 446). Thus, since lexical access takes 100–300 ms to achieve, it is unclear what the role of word "identification" in saccade planning is.

As we saw in Chapter 4, the period after a saccade is initiated is called **saccadic latency**. A distinction which is common in reading models is between the **labile** versus nonlabile stage of saccadic latency. In the **labile stage**, a fixation is started but has not been executed yet, so it is still possible to change it or cancel it altogether. In the nonlabile stage, which follows the labile stage, the fixation is already underway and can no longer be changed or suppressed. Care should be taken not to confuse saccadic suppression (blindness during saccade) and saccadic latency (i.e., no saccade can be planned after one is initiated, past the nonlabile stage).

The Acquisition of Information during Fixations

Research in the **disappearing text paradigm** has shown that reading is unaffected if fixations are masked after 50–60 ms, which means that less than 50–60ms is all that's needed for the processing of the word to get started. This does not, however, mean that words are processed and recognized in 50 ms. On the contrary, the average fixation lasts between 200 and 250 ms, and there is evidence that sentence-level integration of meaning continues across word boundaries (see Contextual Influences section, on the wrap-up effect).

Parafoveal Preview

By "preview" we mean that while the reader is fixating on word *n* and processing it, they are already looking ahead in the parafovea at the next word (and occasionally the one after that as well). The orthographic and phonological processing of words begins during the **parafoveal preview** (i.e., *before* the word is fixated). Indeed, the parafoveal preview can lead to skipping words, since if they are identified there is no need to fixate on them (e.g., function words, such as articles).

Information gathered in parafovea preview includes word length information (word length is important in saccade planning and facilitates reading), orthographic information, (i.e., letter recognition), and phonological information (i.e., it works also for homophones; Pollatsek et al., 1992). Parafoveal preview is generally assumed not to have a semantic content (Rayner & Pollatsek, 2006, p. 617) but there is some evidence that it may (Hyönä & Kaakinen, 2019, pp. 257–8). For example, it causes semantic priming (see Methodologies section). A review of the research on parafoveal processing can be found in Schotter et al. (2012).

Targeting Movements

We will follow Rayner and Pollatsek's (2006) organization of the discussion of targeting movements (i.e., the direction of the saccade) around the two questions: (1) *Where* to look next? and (2) *when* to look next? The two questions are independent of each other and are computed in real time during the current fixation. So, going back to Figure 5.1, when the reader is on fixation 9, for example, they are processing the word "considered" and integrating it with the model of the text they have developed up to that point, but they are *also* previewing parafoveally that the next word is probably "the" and that it can thus be safely skipped, and they are also assessing the rough shape of the word "rightful" (without knowing it is the word "rightful") to target the next saccade roughly to the center of the word.

Where to Fixate Next

The visual system must make a decision of where to direct the next saccade. This is determined largely by low-level visual cues in the text, such as word boundaries (i.e., the spaces between words) and word length, which are acquired during parafoveal preview. This is attested by the fact that loss of information about word length and spaces slows down reading.

Generally, the first fixation on a word falls halfway between the beginning and the middle of the word. This is called the **optimal viewing position.** As the reader will recall, saccades are ballistic and so difficult to control. Sometimes the reader will under or overshoot and will thus have to correct, with a smaller saccade that takes place very quickly after the "improperly" targeted saccade. Most corrective saccades happen after return sweeps.

When to Move?

The decision of when to move is largely dictated by the time it takes to process the word. There are numerous factors that affect processing time. Generally speaking, we can say as a first approximation that processing complexity largely determines the decision to move the eyes. However, as we already discussed, processing is not the only activity taking place: parafoveal previews and the planning of the saccade also take place during the fixation. As we mentioned earlier, fixations may range from 50 ms to 500 ms. The shortest are probably before a corrective saccade, whereas extremely long fixations are usually the sign of extreme complexity or processing trouble (e.g., a syntactic anomaly, such as a garden-path sentence; see the following). When the processing becomes too complex or when the reader wants to check that they understood properly what they read previously, they may make use a backward saccade, that is, a **regression.** For more discussion, see Lexical Influences section.

Regressions

Regressions constitute about 10 percent of all saccades of proficient readers but can go up to 15 percent and of course higher, in the case of problems. Shorter regressions can be caused by oculomotor errors. These are targeting errors that occur while making the saccade. Saccadic planning is not actually very accurate. When the saccade is too short, we may get a very quick saccade forward, to get near to the optimal viewing point. If the eye overshoots the optimal viewing point, a backward saccade (regression) places the eyes closer to the center of the word.

Regressions can also be caused by problems encountered during linguistic processing. These tend to be longer regressions. A good example of a linguistic processing problem that causes regressions are garden-path sentences.

(1) The horse raced past the barn fell

Contrast with "The horse, which was raced past the barn, fell." They are discussed more in depth in Syntactic Influences section.

Skipping

The following characteristics lead to more frequent skipping:

- High-frequency words
- Function words
- Short words
- Highly predictable words

Function words tend to be short and high frequency,[5] and thus the first three reasons are somewhat interconnected. For example, in English some of the most frequent words are articles (*the, a/an*) and forms of the verb to be (*is, was*), which also happen to be short.[6] Function words include the prepositions (*on, to, by, with, etc.*), conjunctions (*and, but, or, etc.*), and particles (*not, to*), which are also high-frequency words and short. The reason for skipping seems to be that the reader has already identified the word during parafoveal preview, and so there is no further need for fixating on it. Rayner and Pollatsek (2006, p. 619) point out that there is some evidence that fixations before a skipped word tend to be longer, that is, include some effort to identify the following word.

This concludes our discussion of targeting movements. The overall picture that emerges is that the reader is not at all passively scanning letters but rather engaging in a continuous give and take with the text extracting information, scanning ahead to make educated guesses as to where the next saccade should land, correcting any problems as they arise, and optimizing reading time by skipping those words that are not necessary. This should answer the question of why proofreading is such a difficult task: in order to be a good proofreader, one has to more or less unteach oneself how to read efficiently!

Influences on Eye Movements in Reading

Lexical Influences

Frequency and Length

The **frequency effect** is one of the most well-established results in the eye tracking of reading. It states that there is a direct positive correlation between frequency and speed of processing; that is, the most frequent words (such as "house," "little," and "said"[7]) are recognized faster than relatively infrequent words such as "soliloquy," "relative," and "infrequent" (Just & Carpenter, 1980; Rayner & Duffy, 1986; Rayner & Pollatsek, 2006). Word length and frequency are strongly correlated: more frequent words are shorter and conversely, less frequent words are longer.[8] Studies have been made to disentangle the two factors and have shown that frequency is a significant factor even when length is kept constant. For example, Rayner and Duffy (1986) used stimuli such as

(2) The exhausted <u>steward</u> left the plane.
The exhausted <u>student</u> left the plane.

where we have underlined the stimulus words (unlike in the experiment). Student has a higher frequency than steward, while being the same length (seven letters).

Predictability

Words that are highly predictable from the preceding context are fixated for less time or skipped altogether. Predictability is defined in terms of **cloze probability.**[9] Generally speaking the cloze test consists of sentences with missing elements. The cloze probability of a word is the likelihood that speakers will finish the sentence with it. For example, in the sentence

(3) In the winter, migratory —— fly south.

the word "birds" would have a very high probability of being used, whereas the word "coconuts" has a probability near zero. For further discussion on predictability, see Staub (2015).

Other Factors

Besides frequency, length, and predictability, we also have a few other phenomena that affect lexical processing.

Familiarity As one would expect, more familiar words are easier to process (Williams & Morris, 2004; Rayner & Pollatsek, 2006, p. 622 and references therein).

Complexity Easier words are easier to process and thus are fixated less. In the **spillover effect** the time required to process a difficult word carries on while the speaker is already fixating on the next one. This has the effect that the fixation time on a word after a low-frequency word is longer. Conversely, fixation of the word prior to a low-frequency word is not affected (Rayner & Duffy, 1986; Reichle et al. 2003, p. 448; Remington et al., 2018). Complexity can take different forms: morphological complexity (see Chapter 4), lexical ambiguity (Rayner & Duffy, 1986, p. 197), or length (see Lexical Influences section).

Age of acquisition A significant effect showing that earlier age of acquisition facilitates lexical processing has been found, even after accounting for frequency. While the issue is complex, as age of acquisition is commingled with several other factors (e.g., concreteness, as obviously words acquired at a younger age will tend to be more concrete) recent research (see Juhasz & Sheridan, 2020 and references therein) points to general plasticity (i.e., the brain is more flexible when learning the first concepts and as it becomes organized it becomes less flexible)[10] and the organization of the semantic network (i.e., the words learned earlier become more "central" in the representation of meaning).

Ambiguity When a word has more than one meaning (and most words do) immediately after perception (around 200 ms) *all* the meanings are activated.[11] However, ambiguous words can be balanced or biased; that is, the meanings may be equally likely to occur (e.g., *pitcher* in which the baseball-related and the vessel meanings are roughly equally frequent), or one may be more likely to occur, for example, *scale* in which the weight-related meaning is more frequent than the fish-related meaning. Ambiguity slows down processing and hence produces longer fixations. However, this is the case only for balanced words (in a neutral context, that does not favor either sense). In the case of a biased word (again, in a neutral context) there is no slowdown. Essentially, what happens is that very quickly (between 100 and 350 ms after activation) a meaning is selected and thus the nonrelevant nonsalient meaning is rejected and hence there is no reason for longer fixations. Finally, we need to consider the situation in which the context is in favor of the less salient meaning of a biased word. In this case, there is a slowdown with longer fixations, because of the conflict between the two senses that needs to be resolved.

Orthographic effects and phonological similarity An interesting lexical effect tied to spelling (orthography) is the **orthographic transparency effect:** opaque spellings slow down processing. By opaque spellings we mean spellings that do not have a straightforward relationship between sounds and spellings whereas transparent orthographic systems tend to have a one-to-one relationship between spelling and phonemic system. Obviously there is a continuum, with some languages such as English and French placing on the opaque side and others, such as Swahili, Spanish, Italian, and Turkish, falling on the transparent side. Our favorite example is the French word for "waters": *eaux* transcribed in IPA as [o] in which *none* of the graphemes (letters) correspond to the phoneme. Within a language there can be words that are more or less opaque, for example, in English the words *eye* (opaque and irregular) and *mood* (opaque but regular).[12]

The connection between spelling and pronunciation of the words and reading speed is well established. Homophones (i.e., two words that sound the same but are spelled differently, such as *night* and *knight*) have longer fixations than unambiguous words. The same is true of homographs, that is, two words spelled the same but pronounced differently, for example, *read*, which may be pronounced as [rEd] (past tense) or [rid] (present tense). Consonants are more salient (cf. fast priming methodology; see Methodologies section) as well as initial and final letters.

Remarkably, eye tracking can even offer information on whether the speakers during *silent* reading actually activate the phonological description of the words. Ashby and Clifton (2005) found that four-syllable words with two stressed syllables took longer to read and showed more regressions than words of the same length with just one stressed syllable, regardless of word frequency. This is taken as evidence that readers build a full phonological representation of the words during reading, including their stress pattern (which is not indicated in writing, in English). Breen and Clifton (2011) show that metrical structure (i.e., the alternation of stressed and unstressed syllables) also affects reading speed when the metrical structure does not match expectations (e.g., in metrical poetry). Syllable structure (i.e., the sequence of consonants and vowels, typically notated as, e.g., CCVC = *spit*) is extracted already during parafoveal preview.

Semantic and morphological influences: Semantic similarity (e.g., the words *king* and *queen*) has been found to have a priming effect (i.e., facilitate reading; see Methodologies section). Conversely, semantic clashes (e.g., using a count noun as mass: *John bought too much car for himself*) and complex morphology (e.g., compounding) slow down reading. Hyönä et al. (2020) argue that this is due to the fact that longer compounds do not fit in the foveal area and thus need to be analyzed morphologically and semantically, which of course takes time. Indeed, in their study shorter compounds barely show any lengthening of reading time.

Other areas of semantic influence are plausibility and anomaly. For example, in Example 4

(4) a. The woman used a knife to cut the tough bread very carefully.
b. The woman used a saw to cut the tough bread very carefully.
c. The woman used a book to teach the tough bread very carefully.

(4.a) is the control (normal) sentence, (4.b) is implausible, and (4.c) is anomalous. Various studies have shown that when readers encounter an anomalous or implausible situation longer fixations ensue. However, the effect is immediate in anomalous words and delayed in implausible ones (Rayner et al., 2004).

Syntactic Influences

As we saw in Regressions section, garden-path sentences are sentences that are syntactically ambiguous along the lines of Example 1, repeated here for ease of reference.

(5) The horse raced past the barn fell.

"Raced" is the V of the VP in the initial reading but it turns out to be inside a modifier clause "[which was] raced past the barn" (second reading). The ambiguity is resolved at some point in the sentence, thus forcing a reinterpretation of the previously parsed material. This reinterpretation causes longer fixations and often regressions. Interestingly, it does not cause a slowing of reading, unlike in lexical ambiguity where we do see a slowdown of reading. This seems to imply that readers are insensitive to syntactic ambiguity (and anecdotally we can add that discerning it is often hard even for students doing a grammatical analysis).

The underlying theory of sentence processing is that the syntactic parser part of grammatical competence does not follow all possible analyses in parallel but rather picks one and follows it (Frazier & Rayner, 1982). Of course if the choice turns out to have been incorrect (e.g., because subsequent information disambiguates the sentence) the parser has to backtrack and reanalyze the sentence. This process is compared to following a garden-path. The implication is that if the parser needs to revise its analysis this will take time and result in longer reading times and possibly actual backtracking of the eyes (regression fixations) to confirm the new interpretation. We will not follow the complexities of the syntactic theories underlying garden-path versus parallel processing, but as Frazier and Rayner (1982) point out it entails that when a reader encounters an ambiguity they use two strategies:

1. Late closure principle: attach any new words to the closest node

   (6) John decided to leave yesterday

   the reading in which leaving takes place yesterday is preferable to the one in which the deciding takes place yesterday.
2. Minimal attachment: prefer the simplest syntactic tree, i.e., fewer nodes.

For an example of the parallel processing approach, see Spivey and Tanenhaus (1998). For a discussion of more recent work, which broadens the perspective to other types of ambiguity, see Clifton and Staub (2008) and Grant et al. (2020). In their view, the preponderance of evidence leans toward the serial, garden-path processing of syntactic ambiguity.

Contextual Influences

It is well established that there is an inverse correlation between contextual predictability and time of fixation; that is, the higher the predictability the shorter the time of fixation. Furthermore, predictable words are skipped more frequently. Predictability is determined by the cloze probability (see Chapter 4). Within contextual influences we can distinguish between transitional probability (i.e., the likelihood that $word_a$ is followed by $word_b$; the frequencies being obtained by a corpus) and contextual predictability: for example, in the following sentences:

(7) a. Mary went wild hog hunting with a high-powered <u>rifle</u>

b. Mary bought herself a <u>rifle</u>

In (7.a) "rifle" is highly predictable, whereas in (7.b) it is not. Thus idioms or formulaic phrases such as *buy the farm* or *nest egg* have high transitional probability, as do collocations such as *strong coffee*, *avocado on toast*. However, contextually an item may be extremely predictable even with a low transitional probability, as in *Early studies in eye tracking used plaster of Paris <u>cups</u>*. The transitional probability of *Paris* to *cups* is presumably extremely low, but in the context of the study of early eye-tracking research, it is highly predictable. Studies have shown that high transitional probability shortens fixation time (McDonald & Shillcock, 2003; Moers et al., 2017). However, further research has shown that this is true of all formulaic expressions. For example, Frisson et al. (2005) showed that predictability alone plays a significant part in the processing advantage of formulaic expressions. Carrol and Conklin (2020) claimed that both predictability and frequency play a role. In idioms, presumably because of their fixed nature, the contextual effect was felt more strongly; for example, they found a much higher rate of skipping of the last word of the idiom. So, generally speaking, we can conclude that idiomatic expressions or more generally multiword expressions are processed as a single lexical unit and hence faster.

Wrap-Up Effect

The wrap-up effect refers to the fact that readers pause at the end of the sentence or clause to integrate its meaning. Studies initiated by Fodor and Bever (1965) using the perception of clicks showed that the syntactic units (clauses, phrases, etc.) have some degree of psychological reality. Just and Carpenter (1980) incorporated this idea in their reading model and, as far as we can determine, introduced the term "sentence wrap-up," which has been quite successful. According to Just and Carpenter (1980) the wrap-up effect takes place at the end of the sentence and involves different functions: the search for referents that have yet to be assigned, the clarification of relationships between clauses, and the resolution of any inconsistencies. By the end of the sentence any ambiguities are usually resolved and a "thought" is concluded (no definition of a thought is given). Inferences connected to lexical items (see Lexical Influences section) are processed at the end of the sentence (p. 346). Furthermore, readers pause at the end of sentences to "integrate" the meaning of what they have just read. Finally, memory for the verbatim wording of a sentence declines immediately after it is processed (but does not disappear completely), which indicates that once the meaning of the sentence is stored in long-term memory it is no longer necessary (or possible) to retain the memory of the exact wording of the sentence. Just and Carpenter estimated the wrap-up duration at about 70 ms. They also pointed out that wrap-up may occur at the end of clauses or paragraphs, and so it is not connected exclusively to the syntactic unit of the sentence. Paragraph-level wrap-up is in the 170 ms range (Just & Carpenter, 1980). The clause and sentence final effect, both for reading times and regressions, was replicated in Rayner et al. (2000) who also noted that after a boundary the saccades tend to be longer. Wrap-up times are also sensitive to the reader's motivation: if a sentence is very relevant to the reader, wrap-up times increase (Hyönä & Kaakinen, 2019, p. 261). Hirotani et al. (2006) showed that punctuation (commas and periods) facilitates the wrap-up effect. If the punctuation is removed, the effect is reduced but the number of regressions increases: "readers do not like to reread material after crossing a punctuation mark and, perhaps for that reason, they dwell before crossing the punctuation mark in the first place" (Hirotani et al., 2006, p. 430). Another explanation, not considered by Hirotani et al., is that the absence of the punctuation makes the processing harder and thus requires more regressions. Stowe et al. (2018), while only partially focused on eye tracking, provided a broad review of the literature on the wrap-up effect.

The idea of a wrap-up effect is not without its critics. Warren et al. (2009) reported on a study showing that integration cannot be the only case of the wrap-up effect and suggest that implicit prosody (Chapter 4) and difficulty associated with punctuation may also contribute to the effect. Stowe et al. (2018) reviewed the significant evidence in favor of immediate processing at the morphological, syntactic, and semantic levels (largely matching the evidence we have discussed in this chapter) and pointed out that immediate processing is problematic for a sentence wrap-up effect. Inferences (see later in the text) and contextual integration also are not delayed until the end of the sentence. The only effect that is reliably seen at the end of a sentence is a reaction in brain activity for semantic ill-formedness (the N400 wave ERP). Thus Stowe et al. (2018) cast significant doubt on the very existence of a wrap-up effect specifically and uniquely found at the end of sentences and clauses.

Methodologies and Measures

We will now review the principal methodologies used in the literature to reach the conclusions discussed so far.

Methodologies

The **moving window paradigm**, originally used in McConkie and Rayner (1975), has been extremely influential. A good review of its uses is Rayner (2014). Essentially, it consists of replacing the letters in a sentence with "x," except for a window around the foveal fixation in which the original letters are left unchanged. This allows the size of the window to be controlled independently while the speed of the movement is determined by the eye motion of the reader: as they fixate on a different word, the window moves along with their gaze. Figure 5.2 is an example of a moving window. The participants would see one line at a time. The moving window paradigm was used to determine the size and asymmetry of the perceptual span.

The **disappearing text paradigm** (Liversedge et al., 2004) is a variant of the moving window paradigm, where the text is not marked by the row of "x" but instead it disappears after 60 ms. When the reader moves on to the next fixation the new fixated word disappears after 60 ms and the previous one reappears. Significantly, the reading process is not affected by the disappearance of the text. This does not mean that the lexical processing is complete within 60 ms, but rather that that is enough time for the process

Moving window paradigm

```
This is a xxxxxxxx xxx xxxxxxxxxxxxx xxxxxxxx.
 *

xxxx xs a sentence fox xxxxxxxxxxxxx xxxxxxxx.
             *

xxxx xx x xxxtence for demonxxxxxxxx xxxxxxxx.
                    *

xxxx xx x xxxxxxxx xor demonstratiox xxxxxxxx.
                           *

xxxx xx x xxxxxxxx xxx xxxxxxxxxtion purposes.
                                       *
```

FIGURE 5.2 *An example of the moving window paradigm. The asterisk and the gray (red in the online version) letter indicate the foveal fixation.*

to be activated. Since fixations last between 200 and 250 ms, on average, the rest of the time must be used to plan and execute the next saccade, while concurrently processing the meaning of the lexical item and previewing information to the right of the currently fixated word (Rayner et al., 1981).

The **priming paradigm** is the adaptation to eye tracking of a well-known technique in psychology. Since Meyer and Schvaneveldt (1971) discovered semantic priming (i.e., that a word "facilitates" the recognition of a second word if they are semantically related), priming has been extended to many other areas and has become an important area of research in psychology. Priming has been used in various ways within eye tracking. For example, Carrol and Slowiaczek (1986) presented their subjects with sentences like

(8) The salesman said that the *cloth* was actually *cotton* which had been dyed.

in which a category term ("cloth") was followed by an example (instance) of it ("cotton"). They found that fixation times on the second word were shorter (i.e., the word was processed faster) if the word was primed by the category than if it was not. In other cases, priming is accomplished by displaying a prime for a very short time (e.g., 30 ms) and then replacing it with the target text, to test what primes facilitate reading.

In another setting, the **boundary paradigm** is combined with the priming paradigm to test for preview effect (parafoveal information). For example, Balota et al. (1985) presented their subjects with sentences like

(9) Since the *wedding* was today, the baker rushed the wedding *cake* to the reception.

except that the reader would in fact be presented with

(10) Since the wedding was today, the baker rushed the wedding —— to the reception.

and the blank would be filled by either "cake" or "pie" or by a nonword that looked like them (e.g., "cake" and "cach," i.e., the nonword). A threshold was set up (e.g., the letter "n" in "wedding") so that when the reader's gaze crossed that threshold the stimulus which had been presented in the slot up to that moment was replaced imperceptibly during a saccade with the correct word (cake or pie). The idea here is that if the readers have access to parafoveal preview the words cake or pie should be primed, albeit to different degrees of strength, since pie is less typical, whereas the nonwords should not result in any priming effect. Indeed, Balota et al. (1985) found that fixation times were shorter if the correct word had been parafoveally previewed and thus primed. Also, as predicted, the effect was stronger with the typical prime (p. 375).

Measures

Along with the methodologies discussed in the previous section, scholars have developed a reading-specific set of measures. We review some of them in what follows, without any attempt at completeness.

Single-Word Measures

These measures apply to individual words. The **first fixation duration** is the length of the first fixation, exclusive of any other ulterior fixations. The **single fixation duration** conversely is the length of fixation of an individual fixation, regardless of the order. **Gaze duration** is the sum of the length of multiple fixations (but not of regressions). In contrast, **total time** is the sum of all fixations, including regressions. Finally, **go past time** is the total time elapsed from the first fixation on a word to the first fixation to its right (i.e., past it), including regressions.

Measures for Areas of Text

These measures apply to regions of the text of more than one word, as defined by the researcher. For example, when looking at garden-path sentences (mentioned in Regressions section and discussed more in depth in Syntactic Influences section), the researcher may want to know how long it takes to get past the word that resolves the ambiguity in the garden-path sentence.

These measures include **first-pass reading time**, that is, the gaze duration on a region; **regression path duration**, that is, go past time applied to a region; and **total reading time**: total time, that is, the sum of all fixations, including regressions, for a region.

Models of Eye Control in Reading

Several models of eye control have been proposed, among which the two best-known and most frequently discussed models are the EZ reading model (Reichle et al., 1998; Reichle et al., 2003) and the SWIFT model (Engbert et al., 2002).

We will review them briefly, but we will also mention the so-called "rational" models. An accessible review of reading models, including eye control, can be found in Rayner and Reichle (2010) and Reichle (2021). The latter is for the more advanced reader.

EZ Reader

In the EZ Reader model, attention is allocated serially to one word at a time. Once the word has been identified and access to the lexical features is about to happen, the planning of the next saccade begins, but attention remains on the word until lexical access is complete. If the word is familiar or predictable from the co-text, recognition of the word is faster.

As can be seen from Figure 5.3, the model consists of a visual system, which receives input from the eye. In the foveal region (maximal visual acuity) we note in the center the location of fixation (in this case, the word "into") while the attentional spotlight (see Chapter 4) is on the next word ("the"). This input is processed in the visual system which provides input to the oculomotor system (responsible for moving the eyes, i.e., the saccades) and to the word identification system. Once a word is identified, attention is moved onto the next word and the oculomotor system is informed to start planning the next saccade. Once the process reaches the nonlabile stage, a saccade is generated and the process restarts from the new position of the eyes.

The EZ Reader model has undergone many revisions, which generally speaking have allowed it to overcome some of its original limitations and to accommodate parafoveal-on-fovea effects, word skipping, and so on.

SWIFT

In the SWIFT model (see Figure 5.4), unlike EZ Reader, attention is allocated for up to four words in parallel (Engbert et al., 2002, 2005). For example, one may be fixating on a given word W_n while still processing word W_{n-1} and previewing word W_{n+1}. However, in the SWIFT model attention is not distributed evenly. Processing speed depends on word position (i.e., n, $n+1$, etc.)

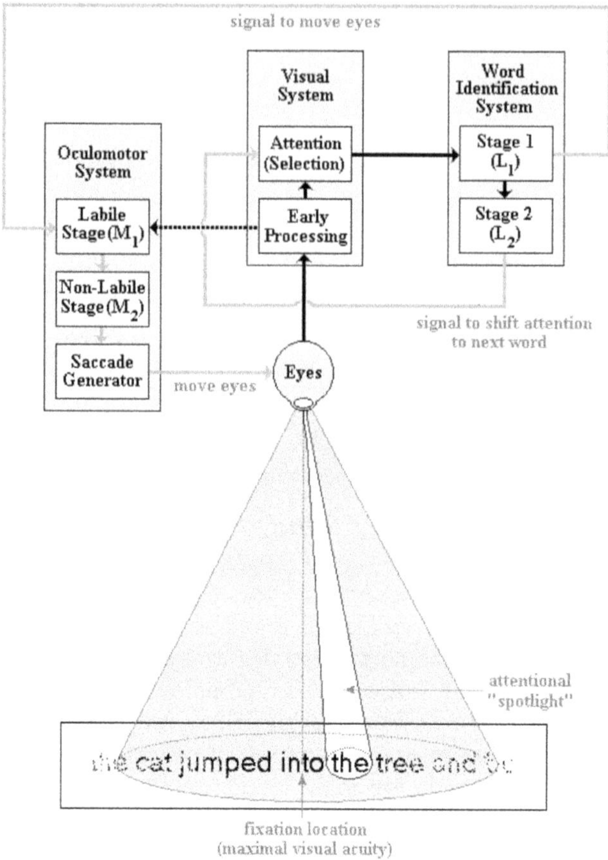

FIGURE 5.3 *The EZ Reader model.* Source: *Reichle et al. (2003).*

and on length of the word so that typically only one or two words are being processed at the same time. Due to the size of the visual acuity field, the range of the words is up to W_{n+2}. Saccades are triggered when the word *n* has been identified and thus if word *n+1* had already been identified, for example, because it is a function word, then a skipping saccade is planned (Engbert & Kliegl, 2011; Risse et al., 2014).

Among the evidence for the SWIFT model are successor effects, that is, the following word (W*n*+1) influences the processing of W_n (successor effects are also known as **parafoveal-on-foveal effects**; e.g., Mirault et al., 2020), and **lag effects**, that is, the processing of word W_{n-1} still affects the processing of word W_n. Regressions are explained as unfinished processing (Engbert & Kliegl, 2011, p. 795).

The foveal inhibition aspect is simply that if there is a high level of lexical processing (in other words, the visual system is trying to figure out what

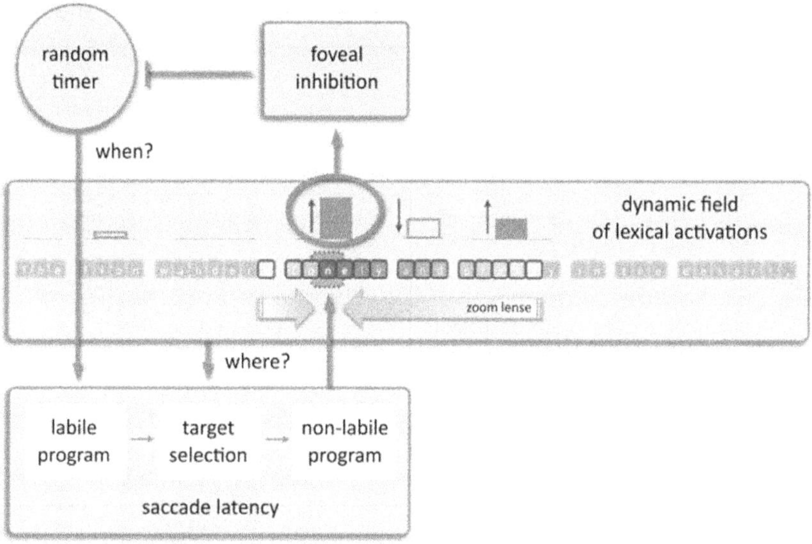

FIGURE 5.4 *The SWIFT Reader model.* Source: *Risse et al. (2014).*

word it is fixating) saccades will be postponed, to some extent. Processing must not necessarily be complete, but the random generator will not fire too soon. The similarities and differences of the EZ Reader and the SWIFT models are compared in Table 5.1.

Rational Models

Rational models are computational. They use simulations to build models that are supposed to match human behavior and attempt to explain their rationality (i.e., show how they minimize effort). They are idealized models, in a sense similar to Chomsky's competence. These models are also called "ideal observers." One may wonder what the purpose of building such abstract models is. Indeed, one of the standard criticisms against rational models is that they are *rational*. Their usefulness comes from the fact that there is substantive agreement between the models and human behavior (Bicknell & Levy, 2010), thus indicating that humans are processing reading information and maximizing efficiency as well.

The first rational model was humorously called Mr. Chips (Legge et al., 1997), after a fictional teacher "with modest intellect" but who is a source of knowledge about the boarding school where he teaches. In short, Mr. Chips is a bit of an *idiot savant.* Mr. Chips has available information on the current input from the "retina" (recall that this is a computer simulation), a lexicon, with associated frequencies, knowledge of the relative accuracy of the saccades

TABLE 5.1 A comparison between the EZ Reader and SWIFT models

EZ Reader	SWIFT
Serial	Parallel
One word at a time	Up to four words at a time
Saccade planned after recognition of word is complete	randomly planned (while recognition is ongoing)
Cannot handle (in its original formulation) parafoveal-on-foveal effects and lag effects	Supported by parafoveal-on-foveal effects and lag effects
Cannot handle skipping saccades	Elegant account of skipping saccades
Target of saccade W_{N+1}	Target of saccade word with highest activation in visual span

it can perform to look ahead or backward in the text, and a principle: "make the saccade that minimizes the uncertainty about the current word." The principle requires a brief explanation, as it is technically worded in terms of entropy or to put it differently as information. Suppose that in the lexicon of Mr. Chips there are only three words that begin with "difficult": "difficult" itself, "difficulty" and "difficulties." Knowing the next letter will reduce the uncertainty to zero: if there is no next letter, the word is "difficult"; if the next letter is "i" the word is the plural "difficulties"; finally, if the next letter is "y" the word is "difficulty." Based on this definition of "uncertainty" and on the other factors, Mr. Chips is able to calculate where the most efficient (i.e., rational) fixation would be. The Mr. Chips model is illustrated in Figure 5.5.

Reichle and Laurent (2006) allowed the program to produce location but and also duration of fixations. Bicknell and Levy (2010) improved Mr. Chips by adding, first, some awareness of context (i.e., the program can "see" more than one character at a time), and, second, the capacity to make a saccade when some degree of uncertainty remains (as humans do). Furthermore, the program can also predict regressions. A development of the Reichle and Laurent (2006) approach is Liu et al. (2013) who used simulations called "artificial reading agents" to determine that a serial processing of words is more common than attending to more than one word. Rational models are used widely, not just in reading but in vision in general.

Conclusion

The study of reading is one of the areas of research in which eye tracking has been deployed most successfully. We believe that our survey of the research,

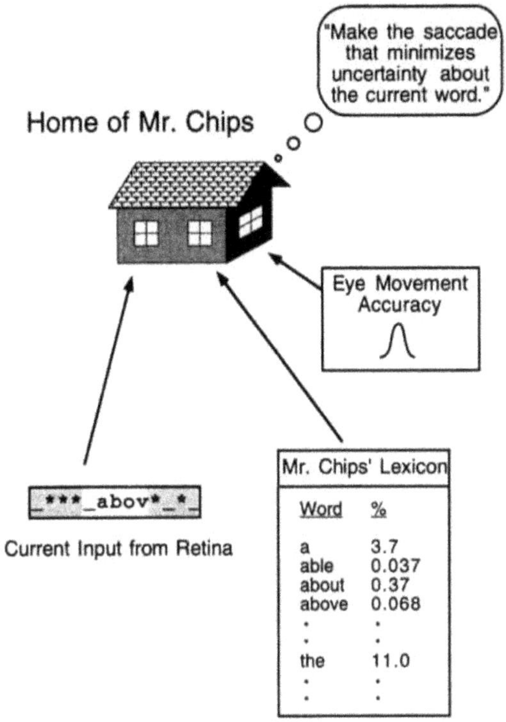

FIGURE 5.5 *The Mr. Chips model.* Source: *Legge et al. (1997), reproduced with permission.*

however inevitably limited and fragmentary, has nonetheless given an idea of the breadth and depth of the achievement of the discipline. The picture of reading that has emerged in the last twenty-five years has become much more detailed and rich than it was before. Finally, the models such as EZ Reader, SWIFT, or Mr. Chips and their further elaborations have set the whole field on a solid theoretical foundation. None of this would have been possible without the research in eye tracking.

Self-Study Questions

The following questions pertain to the contents of this chapter. You should be able to answer them without looking up the answer in the book.

1. Explain why the relatively narrow width of the fovea motivates the saccadic activity involved in reading.
2. What causes regressions?

3. Explain the difference between the visual and the perceptual span.
4. Explain the importance of the parafoveal preview.
5. What is the optimal viewing position?
6. What causes a word to be skipped?
7. Compare and contrast the moving window paradigm, the disappearing text paradigm, and the boundary paradigm.
8. What are the main differences between the EZ Reader and the SWIFT models of reading?

Words to Know

regression saccade	optimal viewing position	return sweep
moving window paradigm	disappearing text paradigm	priming paradigm
boundary paradigm	saccadic latency	labile stage
visual span	perceptual span	parafoveal preview
parafoveal-on-foveal effect	lag effect	

Further Reading and Sources

Most treatments of eye tracking will include a discussion of reading, as it is arguably one of the central concerns of the field. The discussions in Rayner and Pollatsek (2006) and Hyönä and Kaakinen (2019) are very clear.

CHAPTER 6

The Visual World Paradigm

The visual world paradigm (VWP) is one of the great success stories of psycholinguistic research. In a relatively short time span, it has produced an amazing array of studies and results. In what follows we will give a general idea of the methodology and of its potential, but of course we cannot hope to give a comprehensive review of the results and so we will instead focus on some representative studies. In its simplest form the VWP is an experimental technique which provides a visual display, generally but not necessarily divided in quadrants, a verbal stimulus, such as a sentence, and records the gaze of the participants. The underlying assumption is that the gaze is connected to the processing of the stimuli and that, for example, if a dog is mentioned in the text, the participant will look at the dog. This is deceptively simple, because interesting combinations of visual displays and stimuli afford quite interesting results, which ultimately explain its widespread adoption in applied linguistics.

Two Foundational Studies

The historical development of the VWP is unusual, in that the idea of using a board of images and verbal input as stimuli to elicit responses was first implemented by Cooper, a Stanford psychologist, in 1974 and was more or less ignored for the following twenty years, only to be rediscovered by a team of researchers (Tanenhaus, Spivey-Knowlton, Eberhard, and Sedivy) at the University of Rochester, in 1995 when it promptly became one of the most used paradigms in psycholinguistics. Because of the importance of the two studies we will begin by examining them in some detail and will then move on to a review of some of the results achieved within this paradigm.

Cooper's 1974 Pioneering Study

Cooper set out with two goals: first to validate the assumption of the link between gaze and processing and second, assuming the link was found to exist, what it could tell us about language processing. Cooper showed his participants a set of four slides each of which had a 3 × 3 grid of black-and-white drawings (e.g., a lion, a zebra, a snake, etc.; see Figure 6.1) while a short (between 1.5 and 2.5 minutes) reading was being played from a tape. The images were connected to the referents in the readings. The connections were either direct (the word "queen" corresponded to an image of a queen), anaphoric (the word "she" referred to the queen), or indirect (associative: the word "boat" would be heard and an image of a lake would be available on the board). Each reading was followed by a comprehension test.

The outcome of Cooper's study clearly supports the hypothesis that there is a systematic link between gaze and processing:

> When people are simultaneously presented with spoken language and a visual field containing elements semantically related to the informative items of speech, *they tend to spontaneously direct their line of sight to*

FIGURE 6.1 *A reinterpretation of the board in one of Cooper's experiments (Cooper, 1974, p. 93). The images are larger, relative to the board, for better visibility, but the location of the images and type of content matches Cooper's.*

> *those elements which are most closely related to the meaning of the language currently heard.* (Cooper, 1974, p. 85; our emphasis, SA/LP)

What can the presence of the link tell us? The speakers' responses are automatic and anticipatory:

> Ss frequently fixated targets prior to the complete mention of their names or other verbal referents, as anticipatory responses to cues extracted from immediately prior words (e.g. adjectival modifiers), or else from the accumulated phonemes and syllables of the current word (e.g. fixating a zebra, lion, or queen upon hearing *ze*, *li*, or *quee*, respectively). The existence of an anticipatory characteristic implies that people undergo an *active online process of constructing hypotheses* regarding the next successive informative item of speech, and then *use the visual-motor system to test out those hypotheses* prior to confirmation. (p. 104; our emphasis, SA/LP)

Here we see Cooper anticipating a lot of the research that will emerge in the 2000s (see later in the text). However, despite these insights, Cooper's work did not generate any significant following. Huettig et al. (2011, p. 152) reported that only eight citations of the article occur before 1996. Our own searches have found just a few more, but we can safely state that less than twenty citations to Cooper (1974) occur before 1996.

Tanenhaus et al.'s 1995 Study

The goal of Tanenhaus et al. (1995) was squarely to test the modularity hypothesis. The modularity hypothesis (Fodor, 1983) essentially states that the linguistic faculty consists (in part) of modules, each of which is independent from the others. A module accepts input only initially, processes the stimulus, and produces an output. Crucially, a module cannot accept information from another module, except as input. This is all very theoretical, so let's see an example of a modular process: the activation of all the senses of an ambiguous word in the first 100–200 ms: regardless of biasing context, all the meanings are activated.[1] They are quickly suppressed by contextual pressure but they are active in the earliest phases of the processing. Hence, meaning activation is a module, vis-à-vis contextual disambiguation.

In the 1970s–1980s modularity was a central assumption of (certain approaches of) theoretical linguistics. The idea was that syntax generated syntactic structures independently of any nonsyntactic information. These syntactic structures were then overlaid with phonetic and semantic information to form sentences. In other words, syntax was a module. Syntax did not need to know what the subject of a sentence was ("the boy" or "a dog," for example), just that it was a noun phrase that was subject to some kind of modifications and not others.[2]

Tanenhaus et al. (1995) used a 60-Hz head-mounted infrared eye tracker (see Figure 6.2), which according to its maker, Applied Sciences Laboratories, may have been the first infrared head-mounted eye tracker.

The combination of state-of-the-art technology, which allowed more naturalistic collection of data, since it did not require a chin rest, and the direct involvement in the then-central discussion of modularity and syntax no doubt helped popularize the methodology. The Tanenhaus et al. (1995) article is one of the most widely quoted articles in eye tracking in linguistics with over 3,000 citations and definitely one of the most influential pieces of research in the field.

The Tanenhaus et al. (1995) experiment used a physical display divided in four regions, but in the article this is represented by a schematic representation (Figure 6.3); the arrows indicate the direction and order of the movements.

Crucially, the experiment contrasts two presentations: one with one apple and another with two (Figure 6.3). The participants were given two possible instructions, an ambiguous one "Put the apple on the towel in the box" and an unambiguous version "Put the apple that's on the towel in the box." In the two referents ambiguous instruction combination, for example, the participants saw Figure 6.3 and heard "Put the apple on the towel in the box." In this condition, there is an ambiguity: the participants cannot know which apple is meant until the prepositional phrase "in the box" clarifies

FIGURE 6.2 *The eye-tracking equipment used in Tanenhaus et al. (1995).* Source: *Ballard et al. (1995, p. 75).*

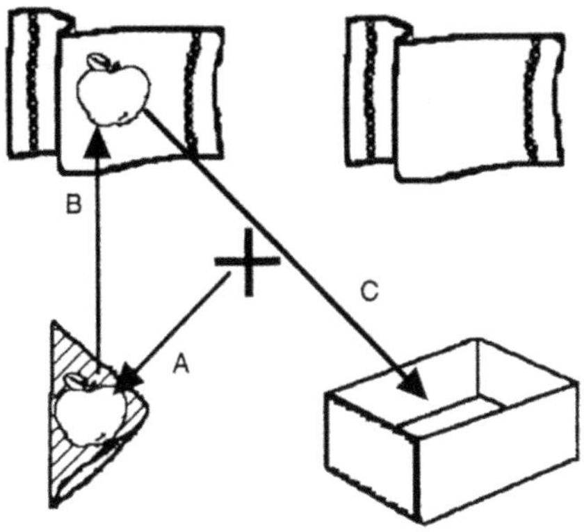

"Put the apple on the towel in the box."

FIGURE 6.3 *The schematic representation of the board used in Tanenhaus et al. (1995). Two referents version.* Source: *Tanenhaus et al. (1995). Reproduced with permission of* Science.

TABLE 6.1 The Tanenhaus et al. (1995) experimental setup

	"on the towel"	"that's on the towel"
1 apple	resolved ambiguity	unambiguous
2 apples	ambiguous	unambiguous

that "on the towel" was a location and not a destination. Summing up, the experiment can be described as a 2 × 2 setup (see Table 6.1).

In the one referent, ambiguous condition 55 percent of the participants looked at the apple, then at the empty towel, and finally at the box, thus showing that they initially understood "on the towel" as a destination (i.e., they were preparing to move the apple to the empty towel). None did so in the unambiguous instruction single referent case. Furthermore, the time required to establish the correct referent in the two referents context was the same regardless of the ambiguity of the prompt, "which indicates that 'on the towel' was immediately interpreted as a modifier, not as a destination" (Tanenhaus et al., 1995, p. 1633). Tanenhaus et al. concluded

> in natural contexts, people seek to establish reference with respect to their behavioral goals during the earliest moments of linguistic processing. Moreover, referentially relevant nonlinguistic information immediately affects the manner in which the linguistic input is initially structured. Given these results, approaches to language comprehension that assign a central role to encapsulated linguistic subsystems are unlikely to prove fruitful. (1995, p. 1634)

In the last sentence by "approaches that assign a central role to encapsulated linguistic subsystems" what is obviously meant is "modularity."

Despite the promising and exciting results in Cooper (1974), the article was largely ignored until it was rediscovered in the wake of Tanenhaus et al. (1995; Huettig et al. 2011, p. 52). What motivated this neglect? And, conversely, what motivated the success of the Tanenhaus et al. paper? We believe it is the result of a convergence of three factors:

1. Cooper (1974) was a psychological study, with no references to linguistics. Tanenhaus et al. (1995), conversely, was full of references to linguistic theory and they used the study to make a point about linguistic theory (modularity). In other words, Tanenhaus and colleagues were fully part of the (psycho)linguistic discussion, whereas Cooper was an outsider to linguistics.
2. Cooper's study used a 3 × 3 board, and his points were about the semantics of sentences; Tanenhaus et al. (1995) used a 2 × 2 board and made points about syntax. So, not only was the Cooper study more complex, it concerned an area of linguistics (the semantic processing of spoken language), which in 1974 was, simply put, outside the topics of interest in the field, whereas Tanenhaus et al. addressed explicitly modularity, a central topic in linguistic theory at the time, as we saw.
3. Huettig et al. (2011) pointed out that the cumbersome nature of the equipment used by Cooper (a glass screen and a video camera that recorded the position of the eyes) and the relatively simpler head-mounted eye tracker used in Tanenhaus et al. (1995) must have affected the different reception. If a scholar wanted to replicate Cooper's methodology, hard work would have been involved, whereas the rapid evolution of eye-tracking technology (see Chapter 1) meant that the technology deployed by Tanenhaus et al. (1995) was within reach of any reasonably well-funded department.

In what follows we will review a sampling of studies that use the visual word paradigm, with the goal of illustrating the versatility of this approach.

Sample Studies using VWP

The VWP can be used in two types of experiments: comprehension and production studies. In comprehension studies the participants are exposed to stimuli and react to them in various ways (e.g., by clicking a mouse on a given image or by reaching for an item). However, generally speaking, their linguistic production is not of interest to the researcher. Conversely, in production studies, the object of analysis are the verbalizations of the participants. In what follows we will exemplify both types of studies.

Processing Studies

Lexical Activation

We know from research in psycholinguistics in the 1970s and 1980s that lexical recognition involves the activation of cohorts of words. A **cohort** means all the words that begin with a certain sequence of phonemes.[3] However, some research had shown that other words that did not share the beginning of the word were also getting activated. So some **continuous-mapping** models were developed, which assumed that, unlike the cohort model in which words "drop out" incrementally if the beginning does not match, the potential competitors are checked continuously; in other words, this allows for activation of words matching at the end (called "rhymes"). Allopenna et al. (1998) used a 5 × 5 grid, with a fixation cross in the center. A sample set of images were arranged in a grid, so as to be immediately above, below, and to both sides of the fixation cross. In one case, the board displayed a speaker, a beetle, a carriage, and a beaker, in a clockwise arrangement. The corresponding words were

- Beaker; the target word
- Speaker; a rhyme
- Beetle; a cohort competitor (note how "beaker" and "beetle" both start with [bi])
- Carriage; unrelated.

They then monitored where the participants looked, as they heard the target word being uttered. As expected, the cohort competitor got as many fixations at the beginning (where the two strings match) but once they started differing the cohort competitor stopped being fixated; however, around 200 ms into the production of the word the rhyme started being fixated as well (although less). The results "clearly show that both cohorts and rhymes compete for lexical activation" (Allopenna et al., 1998, p. 431) and "strongly support" (p. 437) the continuous-activation models.

Influence of Context on Sentence Processing

In addition to lexical processing, sentence processing was also addressed. There are two views of the processing of sentences, one that assumes that syntax is processed first, because it is a module, and then semantic information is "added," and one that assumes that semantic information is accessed immediately along with the syntactic information. If the latter happens to be the case, it would be a strike against the modularity of syntax. Recall now the Tanenhaus et al. (1995) study we reviewed earlier. Since they showed that the visual information has an effect on processing, their study makes the case against the syntax first, modular account.

Chambers et al. (2002, 2004) in similar experiments also showed that semantics and the affordances of objects enter in early syntactic processing, again militating against modular accounts of syntactic processing. Chambers et al. set out to examine whether there is an interplay between the semantics of prepositions and the context which the text is interpreted. They used a visual display which contained either one container (a can) or three containers (a can, a bowl, and a glass). Their stimuli sentences included the prepositions *inside* and *below* (*Put the whistle inside/below the can*). The prediction was that in the case of "below" the interplay of the semantics of the preposition and of the context would be irrelevant, since one can put any object below any other object (especially in a visual display). The same was expected if there were three containers, since the semantics of "inside" does not offer any advantage to any of the three containers. Conversely, the expectation was that in the case of *inside* in the context of a display with only one container, the subjects should start looking at the container earlier, since it was the only one that afforded putting something in it and thus anticipate or predict what the speaker was going to say. This was in fact the result of the experiment. Chambers et al. (2002) concluded that the semantics of prepositions constrain the "dominion of interpretation" of the noun phrase of the preposition and that the affordances of objects do so as well (2004).

Predictive Understanding

Another topic, which, as we saw, Cooper (1974) anticipated is the predictive nature of much processing. This aspect was addressed in Altmann and Kamide's (1999) study. The study was based on a visual display consisting of a naturalistic cartoon of a boy sitting on the floor with a ball, a cake, a toy race car, and a train set on a circular track, all within reach.

During the experiment the subjects were provided with two prompts:

(11) The boy will move the cake

The boy will eat the cake

Altmann and Kamide reported that when the prompt was "eat" eye movements toward the cake started significantly earlier than when the prompt was "move." Since the cake is the only thing that affords being eaten, they concluded that "information extracted at the verb can be used to guide eye movements to whichever object in the visual context satisfies the selectional requirements of the verb. This guidance is initiated before the linguistic expression corresponding to the verb's direct object is encountered" (Altmann & Kamide, 1999, p. 258). In other words, this is evidence that the subjects do not passively wait for information to emerge but they actively anticipate or predict what is coming in the sentence. On prediction in language processing in general, see Huettig (2015).

Pragmatic Inferences (Implicatures)

Of course, VWP studies are not limited to syntax and the processing of lexical items. There have been several studies that have tested the psychological reality of a specific type of pragmatics inferences called **implicatures**. Space prevents us from going into any detail (and interested readers may consult Attardo and Pickering, 2021) but implicatures were introduced by Paul Grice (Grice, 1989) to explain how a sentence may literally state one thing and mean (imply) another thing (i.e., the implicature). Grice introduced the principle of cooperation, which consists of four maxims, which tell the speaker how to formulate their words to be maximally communicative. In what follows we will illustrate some of the maxims and the studies that address them.

Evidence that the maxims are followed comes from a number of studies. For example, the maxim of quantity states roughly that one should provide the right amount of information: too much or too little information is confusing. Sedivy et al. (1999) tested this by having speakers look at a display that contained four objects: two glasses (one of which was taller than the other), a pitcher, and a key. The subjects were instructed to pick up one of the objects and move it. The crucial manipulation of the stimuli was that in some cases only one glass was present (and the second glass was replaced by another unrelated object). The results showed a significant effect: "contrast was also observed, with displays that included a contrasting object resulting in shorter latencies than displays that did not include a contrasting object" (p. 131) or to put it more simply, the expression "tall glass" helped identify the referent faster when there was a shorter glass in the display. This supports the maxim of quantity. It also points at an incremental immediate integration of semantic and contextual information. Other studies that support the psychological reality of implicatures include: O'Brien et al. (1988); Engelhardt et al. (2006; maxim of quantity); Huang and Snedeker (2009); Tavano and Kaiser (2009); Grodner and Sedivy (2011); Degen and Tanenhaus (2015); and Sun and Breheny (2019, with mixed results).

Dialog

In a conversation, speakers establish a shared **common ground** (you can think of it as a set of information that both speakers are mutually aware of; Clark,

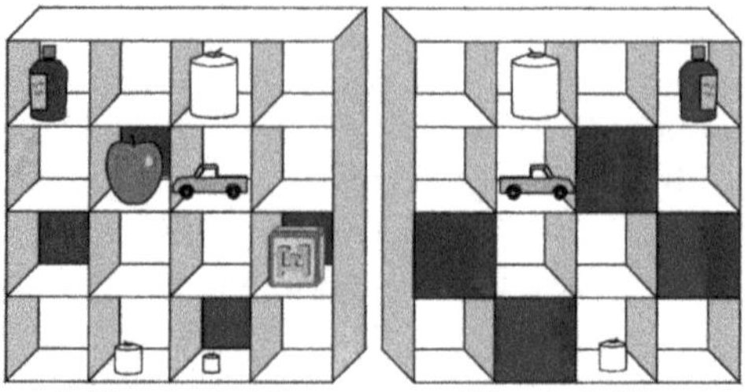

FIGURE 6.4 *The visual display used in Keysar et al. (2000). Reproduced with permission of* Psychological Science.

1996). An important area of research in the VWP has been precisely to assess the effects of common ground in dialogs. Ultimately, the experimental setup is derived, in most cases, from a study by Keysar et al. (2000). They displayed the 4 × 4 board shown in Figure 6.4. The best way to visualize this is a bookcase, with a backing only in the squares with a black back. The participants (addressee and director) were located on each side of the bookcase. Obviously, only the addressee knows what is in the squares with the black back, since the objects are not visible to the director. Unbeknownst to the participants the director was a confederate,[4] who followed a script. The point of the setup was that the addressee and the director have different levels of information and thus they could not assume that it is all in the common ground.

For example, in the addressee's view there were three candles, whereas the director could only see two of them. If the director said "move the small candle" the addressee could take either a common ground reading, and reason that since only the medium-sized candle is visible to the director they must mean that, or an egocentric interpretation and move the smallest candle. Generally, speakers will focus on areas that they know are part of the common ground. There was a significant increase in fixation times when the addressees needed to exclude the referents not part of the common ground. However, in some cases (about 23 percent) the speakers use egocentric strategies (in other words, they fail to take into account the common ground). Keysar et al. (2000) did not take a specific stance on whether the addressees start from an egocentric perspective and work out the common ground in a successive phase or if common ground is active from the beginning of the interpretation, but in Keysar et al. (1998) they seemed to endorse the idea that interpretation starts from the egocentric view and is later corrected as needed. Hanna et al. (2003) using a similar methodology, documented that the common ground effect can happen very close to the start of the interpretation. Barr and Keysar (2006) provided a broad survey of coordination between speakers in dialog, including common ground.

Production Tasks

The same coupling between gaze and attention that underlies reception studies holds for production as well. Meyer et al. (1998) studied speakers engaged in a naming task. They found that "the viewing times depended systematically on the time needed to retrieve the phonological form of the object names" (p. B32) [*sic*]; in other words, the speakers activate the phonological information of the word after recognizing the object before they move on to the next object.

Where does sentence planning start from? Griffin and Bock (2000) showed that, when presented with an image about which to talk, speakers start out by apprehending it, that is, extracting the gist of the image (see Chapter 9, on the concept of gist in visual processing) and then proceed incrementally from there. Furthermore, they fixate on the things they are about to mention, in the order of mention. There was a gap of about 900 ms between fixation and mention. According to Griffin (2004) speakers do not gaze at all the referents in an utterance; rather, they gaze primarily at the referent when selecting names and activating their phonology (pronunciation). These are called "name-related gazes" and are significantly longer than when people gaze at an object for recognition (without talking about it).

While early studies, such as Griffin and Bock (2000), favored a clearly distinct apprehension phase followed by an incremental production phase, later studies have significantly tempered this position. Bock et al. (2004) concluded an extensive review of the literature by stating that there is "a strong tendency for conceptually prominent referents to take the lead in utterances" (p. 259). A number of studies have shown that salient referents tend to be named first and end up as sentential subjects (e.g., Gleitman et al., 2007); that is, they are "aboutees" (Bock & Ferreira, 2014). Gleitman et al. (2007) showed that participants look at perceptually salient characters within 200 ms and start their sentence with it.

Obviously, sentence planning is more complex than naming an individual item, but we should not think that speakers plan the whole sentence before starting to speak. There is clear evidence that the process is "highly incremental" (Huettig et al., 2011, p. 165) and the units involved are fairly small. For example, Brown-Schmidt and Konopka (2008) consider the planning for modification by an adjective such as "small/large." They considered English and Spanish: in English most adjectives are premodifiers,[5] whereas in Spanish they are postmodifiers (again, generally speaking). The experimental setup was a grid, like the one in Keysar et al. (2000) in which two candles of different sizes, for example, were visible. They showed that English speakers, who had less time to make the decision, since the adjective comes before the noun, looked earlier than the Spanish speakers at the contrasting elements. Aside from the incremental nature of the planning process, the units being planned here were smaller than noun phrases and at most that size. Larger units are also planned for, as one would expect. Bock and Ferreira (2014) discuss how syntax mediates between saliency and the conceptual frames activated by the words. There is some evidence that verbs may have a privileged status in the encoding process (see Konopka, 2019 for discussion).

In conclusion, utterance production can also be fruitfully studied with eye tracking and in the VWP paradigm, but for a number of reasons, the topic has received less attention than language comprehension. Nonetheless, we can say that a fairly clear picture of a largely incremental process that starts with a focus on the subject and the verb is beginning to coalesce. Here we would be remiss if we did not mention the words of caution of Huettig et al. (2011, p. 165) that

> the eye movement data may be somewhat misleading because the object a person is fixating upon need not be the only object they are attending to. Instead viewers can use a broader attentional focus and attend to several objects in parallel (. . .) Speakers can attend to two objects—a foveated object they are about to name and an extrafoveal object they will name next—in parallel. Moreover, the processing of the extrafoveal object can be sufficient for its name to be activated in parallel with that of the foveated object. However, parallel object processing only occurs when both objects are relatively easy to recognize and name. (Huettig et al., 2011, p. 165)

Good surveys of sentence production are Meyer (2004), more focused on nouns and noun phrases, Bock and Ferreira (2014), and Konopka and Brown-Schmidt (2014).

Prosody Studies

The kind of predictive interpretation we saw in Processing Studies section is also seen in studies concerning prosody. In the VWP, studies of prosody focus primarily on the prosodic feature of sentences provided to the participants in tandem with visual displays. In what follows we review a few studies, again, just to give a general idea of the kind of topics that can be examined using the VWP.

Contrastive Stress

Contrastive stress affects salience of objects in a visual display. Sedivy et al. (1995) showed that in a board containing simple geometrical figures of various colors, the contrastive stress[6] disambiguates between two alternatives, as in

(12) Touch the LARGE red square.

in a board containing two red cardboard squares, one of which is larger than the other, a large blue circle, and a small yellow triangle.[7] Indeed, participants looked at the target geometrical figure about 100 ms faster, such as in Example 12. This demonstrates that speakers use prosodic information, such as contrastive stress, to resolve referential ambiguities and that the phenomenon is amenable to a VWP treatment.

Kamide et al. (2003) had shown that speakers in German gaze at anticipated referents when they can predict reference to them syntactically. Knoeferle et al. (2005) expanded this to visual disambiguation and prosody. German grammar allows for sentences of the form:

(13) The princess paints the pirate

which are ambiguous (due to the relatively free word order of German) between two interpretations: one in which the princess is the subject of the sentence (i.e., is doing the painting) and one in which she is the model (the pirate is doing the painting). Normally, case markers for subject and direct object would disambiguate, but in some words, such as princess and pirate, the case markings are absent. In the Knoeferle et al. (2005) paper, the participants were shown pictures in which the princess (for example) was holding a painter's palette and thus the image disambiguated the sentence. The study shows that disambiguation is faster (the subject looks at the correct character) when the visual information is available.

Weber et al. (2006) further expanded these findings to prediction via intonation: they found that even though the sentences contained no disambiguating case markers, they hearers gazed at the referent in the anticipated case depending on the intonation of the sentence.[8]

Another interesting factor in predictive/disambiguating phenomena is the effect of **disfluencies**.[9] Arnold et al. (2003) showed that disfluencies precede new referents and that hearers, when hearing fluent speech, anticipate known information in initial position (as seen in Dahan et al., 2002) but when hearing disfluent speech they anticipate new information. Arnold et al. (2004) also found that this applies for an elongated form of "the" ([de] vs. [di:]), which is also a disfluency. See also for further references and discussion Huettig et al. (2011, pp. 158–9).

Finally, it is worth noting that it has been established that the facilitative and anticipatory effects discussed earlier need not appear in a Lx situation, or may appear differently. For example, Ito et al. (2018) show that predictions that take place in L1 may not take place in Lx. Perdomo and Kaan (2019) find that Lx speakers are less able to make predictions. Qin et al. (2019) compare L1 and Lx speakers of Chinese and find that the facilitation effect of tone for Mandarin speakers differs in Lx speakers.

On linguistic prediction in general, see Huettig (2015), Pickering and Gambi (2018), and Huettig and Guerra (2019), who argue, on the basis of the finding that prediction is curtailed under certain conditions, that it is less widespread than commonly believed. Huettig and Janse (2016) find that individual differences, particularly in working memory, affect anticipatory processing.

Interplay between Visual and Linguistic Systems

As we pointed out, the VWP begs the question of the interplay of the visual and verbal systems in the brain. Linguistic stimuli are inherently multimodal in their natural discursive setting. Generally, the verbal (auditory) and visual signals integrate without trouble. However, in some cases the integration fails or is only partial. Hence, we will briefly discuss the phenomenon of interference. In a sense, interference is the opposite of priming. As we saw in Chapter 5, priming facilitates the retrieval of a lexical item, for example, whereas interference slows down or prevents integration.

The Stroop Effect

In its simplest form the **Stroop effect**[10] consists of a naming task in which words are presented to the speaker printed in colored ink. The color of the ink does not match the word, that is, it is incongruent (e.g., the word *red* printed in green ink).[11] The task is surprisingly difficult to manage (it is easy to try at home). The increased cognitive load of the Stroop effect has been documented in pupillometry (Laeng et al., 2011; see Chapter 4).

The Stroop effect is used as a test (the Stroop test) to assess executive control, that is, the capacity of the brain to monitor some of its processes. Executive control turns out to be very important in processing syntactic ambiguity, such as garden-path sentences, for example (see Chapter 8 for further discussion). One of the common uses of the Stroop effect in linguistics is to assess the "bilingual advantage" (e.g., Bialystock et al., 2004): since bilinguals have to consistently monitor to avoid interferences across languages, it is widely assumed that they develop greater executive control and hence they would do better on the Stroop task. Kousaie and Phillips (2012) found no evidence of bilingual advantage among Canadian French-English bilinguals. Singh and Mishra (2013) found evidence of a bilingual advantage but found that it correlated with proficiency: higher-proficiency Hindi-English bilinguals outperformed low-proficiency bilinguals (see also Godfroid, 2020, p. 102) on the Spatial troop task (see later in the text). Proficiency and other factors (such as script) are examined in Coderre et al. (2013) using English-Chinese and Chinese-English bilinguals; they also found evidence of a bilingual advantage.

There are numerous variants of the Stroop effect: for example, an emotional Stroop, in which color words are replaced by words with associations with depression or neutral words, a spatial Stroop with arrows pointing up or down, a numerical Stroop (with numbers of different size), and more. They are used for various psychological tests.

The McGurk Effect

A different, but equally interesting case of interference between auditory and visual channel is the McGurk effect. The facial movements, specifically the movements of the mouth, can have an effect on what one *hears*. If the syllable [ba] is played with the video replaced by the lip movements of [ga], speakers reported hearing [da]. This is known as the **McGurk effect** (McGurk & MacDonald, 1976).[12] The McGurk effect is ideal to highlight the integration of the visual and auditory channels: when there is interference between the two one hears a different sound (the fusion of the two stimuli), not an attenuated version of one or two sounds overlapping. In other words, the brain elaborates a perception from all the information it has available. The effect is completely sensitive on seeing the movements of the mouth, even if just in peripheral vision (Paré et al., 2003, p. 565; Hessels, 2020). Both the Stroop and the McGurk effects are automatic; that is, they are not controlled volitionally and persist even when one is aware of the nature of the phenomenon (i.e., knowing one is actually hearing [ba] does not change the fact that one subjectively hears [da]). The effect is felt also for full words and sentences (Wright & Wareham, 2005). This has particular significance in forensic settings in which the integration of visual and auditory effects may be only partial.

The Basic Components of a VWP Study

As we saw the accomplishments of the VWP are quite outstanding, especially considering the relatively short time span of its development, the quarter of century from 1995 when the method was rediscovered to the present day. No doubt this success is in part due to the versatility of the experimental setup. In this section we will very briefly sum up the various components of a VWP study, with the intention to show the versatility of the approach.

1. The type of display may vary. Displays can be in the physical world (as was the case in Tanenhaus et al., 1995), realistic visual scenes, grids with drawings (e.g., Allopenna et al., 1998), or words on a grid (Salverda & Tanenhaus, 2010). For a discussion of an example of study using a grid of words, see Chapter 8. In some

studies the visuals are removed during the response phase ("blank screen studies," see Altmann, 2011, pp. 987–8).

2. The verbal stimulus—the participants in a study may or may not hear a verbal stimulus. In other cases, the stimulus will be delivered in writing.
3. The task may vary significantly and as the reader will recall from the early studies of Buswell and Yarbus (see Chapter 1) the nature of the task affects the gaze patterns. Tasks include finding an item, clicking a mouse, answering a question, or merely looking at a board.
4. The participants may be tested individually or in pairs (in dialog studies). The participants may be engaged in a shared task.
5. The results and their statistical analysis.
6. The VWP allows for both production and comprehension studies. In production studies no verbal stimulus is provided by the experimenter; instead the participants' task is to verbalize.
7. The approach lends itself to examining all sorts of topics, some related to the issue of referent identification and modularity and others having to do with pragmatics, discourse, and even prosody.

The Linking Paradigm

As we saw, the VWP is predicated on the assumption that there is a link between vision (fixation) and language. There are several ways that this linking may take place. They can be summed up as follows:

1. language and vision are independent, processed in parallel and integrated;
2. vision and semantics are activated prior to lexical items (i.e., first we see things and what they are, then we name them);
3. mental world representation: as in (1) but inclusive of the context and affordances dictated by it;
4. patterns of activation of phonological, semantic and visual features are associated by the experience of vision; there is no lexicon, per se (Magnuson, 2019).

However, the very assumption of the eye-mind link has been questioned. Orquin and Holmqvist (2018) pointed out that there is no direct one-to-one mapping between gaze and attention. Indeed, covert attention and phenomena such as inattentional blindness clearly show that there can a significant decoupling of gaze and attention (pp. 1651–3). So, in their view, the correspondence between the two should be proven not assumed. Huettig et al. (2011) pointed out another potential limitation of the VWP: ultimately all that the VWP can tell us is what speakers do when they are looking at pictures (or have been looking at pictures). However, not all language use takes place with a visible referent. Thus to some extent, we still need to prove that speakers behave the same way when there are no visual referents.

Conclusion

In its relatively short lifespan, the VWP has achieved nothing short of a prodigious expansion. It has proven to be a reliable and useful methodology, enabled by eye tracking, with applications across the span of linguistics, often providing evidence in areas, such as language production, where linguistics had been reduced to speculation for over a century.

Not only the breadth of the applications of VWP is remarkable, but the depth in which some topics, such as the resolution of syntactic ambiguity, lexical activation, and so on, have been pursued is remarkable and affords the integration of the VWP with other research techniques such as ERP, fMRI, and EEG, which shows a degree of maturity of the field. Its impact has been "transformative" (Magnuson, 2019, p. 114) on psycholinguistics.

This is not to say that the VWP is without its problems. As Huettig et al. put it, "What these studies show is what listeners can do, not what they actually do in other contexts" (Huettig et al., 2011, p. 156). In conclusion a balanced assessment is presented by Huettig et al. (2011):

> The visual world paradigm can provide information about message generation and the formulation of utterances about the visual world, but we do not know how similar these processes are to those occurring when speakers do not talk about the visual world but their thoughts and feelings. (. . .) Little is known about the properties of the listeners' representation of the speech input, and it is difficult to say how they might differ from representations they build up when they are presented with auditory information alone. (Huettig et al., 2011, p. 166)

Self-Study Questions

The following questions pertain to the contents of this chapter. You should be able to answer them without looking up the answer in the book.

1. What specific aspects differentiate experiments in the visual world paradigm from other experimental eye-tracking studies?

2. Explain the argument against modularity in Tanenhaus et al. (1995).
3. Explain the significance of the Allopenna et al. (1998) study.
4. Explain "predictive" understanding.
5. Show how VWP studies support Grice's maxims.
6. Explain how VWP studies support the idea of common ground.
7. What are the main points established by production studies in VWP?
8. What does the VWP have to contribute to prosody?
9. What do the Stroop and McGurk effects have in common?

Words to Know

The following is a list of words that have been introduced or discussed prominently in this chapter and that you should be able to define. You may test yourself on them, for example, by making flash cards.

visual world paradigm	modularity	continuous-mapping	cohort
implicature	common ground	disfluency	Stroop effect
McGurk effect	contrastive stress	name-related gaze	rhymes

Further Reading and Sources

Good discussions of VWP can be found in Huettig et al. (2011); Spivey and Huette (2016); Knoeferle et al. (2016); Conklin et al. (2018); Magnuson (2019); and Godfroid (2020).

CHAPTER 7

Social Eye Tracking

Social eye tracking involves the observation and analysis of the interaction between more than one participant. If there are two participants this is called a dyadic interaction and if there are more, a polyadic[1] one. Generally speaking, social eye-tracking studies require at least two recording devices (video cameras or eye trackers) capable of capturing the gaze of both participants. In some cases, while there is are more than one participant, the gaze of only one of them is recorded/analyzed. Pfeiffer et al. (2013) reviewed various methodologies used to study "**social gaze**" (i.e., joint and **shared attention**), including dual eye tracking, **embodied agents**, and virtual environments (see Embodied Conversational Agents section). They also considered fMRI, EEG, and other methods that are outside the purview of this book.

Brône and Oben (2018) articulated what could be considered the manifesto of social eye tracking:

> gaze should be perceived as bodily action invested with meaning, both on the part of speakers and their addressees, relevant for the joint action that is human-human (or even human-computer) interaction. (p. 5)

However, the meaning of which the bodily action is invested is contextually based—not just in the banal sense that actions (and thus communicative actions) occur in context. What is of interest here is not any social interaction per se but rather a subset of the behaviors demonstrated by the participants, which are relevant to the gaze behaviors and, on the other hand, the gaze patterns as they occur *within* other actions. For example, we know that looking at someone else while approaching them is generally considered preparatory to starting a face-to-face conversation (Kidwell, 2014) and so obviously *not* looking at someone is a sign that one is not particularly willing to engage in a conversation or possibly that something is wrong.

Thus the acts of gazing acquire different meanings/interpretations as they occur within the context of social acts (see the discussion of Kidwell's work on children's gaze, later in the text).

Only a full analysis of the gaze, the situation, and the context will tell us what the gaze-relevant items are and what the gazes (or nongazes) mean. Indeed, what is *not* looked at may be as significant as what is looked at: consider the phenomenon of gaze aversion (see Gaze Aversion section), which can be defined as deliberately *not* making **eye contact.**

Origins of Social Eye Tracking

Social eye tracking can be seen as the convergence of four separate paradigms: cognitive neuroscience, the discourse analysis paradigm, work on HCI, and finally the introduction of wearable eye trackers and of automatic recognition of images and gaze. The cognitive neuroscience field is itself the synthesis (Argyle & Cook, 1976) of related strands of research in the neuroanatomy of the brain, ethology, primatology, psychology, eye tracking, child development, and the studies of social impairments, such as autism. Of late, these strands of research have been integrated in a computation framework that involves virtual agents and computer simulations (e.g., Wilms, 2010; Pfeiffer et al., 2012).

Discourse analysis is a method of analysis developed to provide extremely detailed analyses of discursive interactions, similar in this sense to microethnography. Conversations among people are obviously and necessarily multimodal affairs, not only because of the gestures and facial expressions that accompany language, or of the prosodic component of discourse (e.g., tone of voice), or the information conveyed by posture, distance between the speakers (proxemics), and the way they arrange themselves in space (e.g., in small groups, or all facing in one direction), but because fundamentally conversations occur in a context (i.e., reality) and reality is, inescapably, multimodal.

HCI is a broad field encompassing all interfaces between human and computer. In this context, we are only interested in the uses of gaze as part of the interface (usually to determine where the user's attention is directed). Image recognition is also a vast field of computer science and applied mathematics. Here too, we are concerned with some specific applications. As the reader will recall, we discussed the evolution of the eye-tracking technology in Chapter 1. We will now briefly review the paradigms that converge in social eye tracking.

Social Gaze

The cognitive neuroscience approach to the study of "social gaze" is concerned with the study of the communicative uses of gaze and attention, particularly in their phylogenetic origins (i.e., their evolution and development). Argyle and Cook (1976) is a fundamental synthesis of the early phase of the research on gaze. Emery (2000, p. 583) noted that the concept of "gaze" used by ethologists is broader than the direction of the foveal region and encompasses the direction of the head and the orientation of the body. If the eyes are not visible, we can still figure out roughly where a person is looking from the orientation of their face and body. Emery provided a helpful set of definitions, which we will follow. Mutual gaze is defined as two subjects looking at each other ("eye contact"). Averted gaze is defined as one of the two subjects looking away ("averting their eyes"). **Gaze following** is when one subject is aware that the other subject is not looking at them and they look in the same direction as the other subject (they follow their gaze). **Joint attention** is defined as gaze following with an object of focus (both subjects are focused on the object and one is aware of both foci). Shared attention is defined as joint attention in which both subjects are mutually aware of their attention being directed at an object (p. 590).

Humans have the capacity to attribute mental states to other agents (this is known as the "theory of mind"), and gaze direction is an important cue in this process. Mundy and Newell (2007) and Kim and Mundy (2012) distinguished between responding to and initiating joint attention. Responding implies that a child follows the line of sight of an adult, whereas initiating joint attention implies a desire to share an interest or experience with others. Studies have shown that joint attention in infants is associated with mental development, language acquisition, and social competence. Autism has been shown to be associated with impairment of initiation of joint attention but not of responses to it. Primates have been shown to be incapable of initiating joint attention. Schilbach et al. (2010) in a study combining fMRI and eye tracking showed that joint attention activates the parts of the brain related to the processing of communicative intention ("meeting of the minds") and, interestingly, the areas of the brain associated with rewards, suggesting that sharing one's attention is pleasurable. See also Sheperd (2010) for a survey of the research in the area. Now that we have outlined the social gaze paradigm, we turn to the second component of social eye tracking, ethnomethodological discourse analysis.

The Ethnomethodological Paradigm

By "ethnomethodological paradigm" we refer not just to Garfinkel's work (e.g., Garfinkel, 1967) but also to discourse analysis and the conversation analysis paradigm developed by Harvey Sacks, Emanuel Schegloff, and Gail Jefferson in the 1970s. This is not the place to provide an introduction to conversation/discourse analysis, and excellent ones are available, such as Wooffitt (2005) or Sidnell and Stivers (2012), so we will limit ourselves to a few broad strokes to be able to situate social eye tracking within the paradigm.

Conversation analysis emphasizes extremely detailed transcriptions and subsequent analysis of interactions. This attention to detail and emphasis on the small regularities of social practices was very influential. Indeed, Gumperz and Hymes define **ethnomethodology** as incorporating and expanding conversation analysis. In particular, they brought attention to the intentionality of the participants and the broad sociocultural context of the interaction (see, e.g., Hymes's far-reaching SPEAKING model, Hymes, 1972).

Of particular relevance to social interaction is Gumperz's idea of contextualization cues. A **contextualization cue** is any linguistic feature that is associated with a given use of language in context. For example, code switching may signal a switch to an intimate or to a formal mode, depending on the direction of the code switching (from formal to informal, or vice versa). Significantly, Gumperz calls these "cues" and not signs, in the sense that contextualization cues do not *mean* directly, the way lexical items, sentences, or **implicatures** mean[2] in a text, but they merely serve to direct the attention of the speakers to some feature of the context. As we will see, many aspects of gaze are best understood as contextualization cues.

Finally, we should mention that a relatively recent extension of discourse analysis, multimodal discourse analysis (Kress & Van Leeuwen, 2001; Levine & Scollon, 2004; Erickson, 2004), thematizes the fundamental multimodality of all communication and attempts to address all its complexities. This is definitely a growing area of research. Erickson (2011) provided an overview of the development of the uses of video in (micro-)ethnographic work, primarily in the United States. For a broad perspective, see the second edition of the *Routledge Hanbook of Multimodal Analysis* (Jewitt, 2014). The second edition is much expanded from the first one, including, tellingly, a chapter on eye tracking in multimodal analysis (Holsanova, 2014), among others. Further treatment of multimedia learning and eye tracking can be found in Chapter 8.

Human-Computer Interaction

The field of human-computer interaction (HCI) studies was a significant factor in the development of social eye tracking, because the various systems built to take advantage of the gaze of the human and/or to simulate a realistic "gaze" by the (embodied) agent (see Collaborative Tasks section) were inherently dual, at least if one counts the machine. Significantly, the earliest uses of dual or social eye tracking are found in this area.

The Technological Factor

Under this heading we consider two components: the automation of facial recognition and its annotation. The last factor in social eye tracking is not a research tradition or a field of inquiry. It is simply a recognition that the evolution of two unrelated technological fields had to be sufficiently advanced to make the kind of research practiced in social eye tracking feasible: (1) wearable or at least unobtrusive, flexible eye-tracking equipment (see Chapter 1) and (2) automatic transcription and annotation of corpora, including image and facial recognition.

As Carletta et al. (2010), Clark and Gergle (2011), and Brône and Oben noted (2018, pp. 8–9), given the extremely intensive nature of annotating data at the granularity of a hundredth of a second and the need to correlate verbal, gestural, and gaze data, hand coding is not really practical. It has been done, but it is difficult, requires trained raters, and is expensive (as a consequence of the two previous reasons). It is thus necessary to automate at least some of the process to make the approach viable. Since the 2010s, such tools have become available (see a list of tools and discussion in Brône and Oben, 2018), and this may partially explain the dynamic nature of this subfield. A good example of these methods is Pfeiffer and Renner (2018) whose software recognizes gestures and gaze and annotates it with gaze information from the eye tracker, in real time. *Openface* (Baltrusaitis et al., 2018) is an open system that specializes in recognizing facial expressions, based on the Facial Action Coding System (FACS; Ekman & Friesen, 1978). De Beugner et al. (2018) integrate automatic detection of object, faces, and bodies and overlay gaze information on these data in ELAN,[3] with a very high (>97 percent) degree of agreement with a human rater. Schäfer et al. (2020) present a system (descriptively named *MutualEyeContact*) that can extract facial information and mutual gaze from dual eye-tracking data. The system is based on Baltrusaitis et al.'s *OpenFace*.

Dyadic and Polyadic Interactions

We can now introduce the first issue of social eye tracking: the number of participants. For historical reasons, having to do with the technology (see Chapter 1) of eye tracking and its deployment, eye tracking was applied almost exclusively to single subjects. With the introduction of more portable equipment it became possible to observe and record two subjects simultaneously; from there the step to multiparty encounters was short.

Dyadic interactions involve two people (or possibly a person and a virtual agent[4]); **triadic** interactions involve two people and an object (e.g., a computer screen) that both people are directing their gaze at. An interaction with more than two people participating is a **polyadic** interaction. Interactions are based on communication, and gaze is a significant part of any verbal communication. Indeed, some studies have found that it is harder to understand speech without visual cues (particularly, seeing the lips; Argyle & Cook, 1976, p. 158). Moreover, speakers who are using a nonvisual medium, such as the telephone, nonetheless continue to produce gestures regardless of the fact that they cannot be seen (Bavelas et al., 2008). Incidentally, the experimental setup excludes the interpretation that producing gestures that won't be seen is just a matter of habit: in a control condition speakers were asked to produce a monologue recorded for a tape recorder. In this case, a significantly smaller number of gestures was produced, thus showing that the relevant factor is the dialogic nature of the conversation, not the habitual nature of the gestures.

Functions of Gaze

This section briefly reviews the various social functions of gaze superimposed on the physiological function of seeing. We will first examine the functions of gaze not directly linked to speaking and then we will examine gaze functions that accompany speech.

Gaze as Deictic

At its most basic social aspect, gaze directs attention; it is "deictic" (i.e., a pointer). Friesen and Kingstone (1998) showed that gaze triggers a reflex shift of covert attention to the direction of the gaze. This effect has been firmly established; see Shepherd (2010) for a review of the evidence. It seems that this deictic function is the foundation of the theory of mind and that a deficiency in this aspect leads to the manifestation of symptoms of autism spectrum disorder (Shepherd, 2010, p. 1; Grossman, 2017). Some animals are able to follow gaze; for example, great apes and horses can follow human gaze. Dogs can both follow and *direct* gaze (i.e., they can get a human to look at what they are looking at). This seems related to domestication, as human-raised wolves are less able to do so (Shepherd, 2010).

One of the deictic functions of gaze is disambiguation. For example, Hanna and Brennan (2007) established that gaze is used to disambiguate referential expressions, so, for example, the expression "the blue circle with five dots" may have appeared with a visual display in which there were two blue circles, one to the left and one to the right, so that the direction of the gaze would have disambiguated between which blue circle was meant. Moreover, they found that this happens fairly early on in the processing of the sentence, which points to the conclusion that "eye gaze has an automatic, reflexive orienting effect on attention" (p. 613; note that "reflexive" is here used in the meaning of "as a reflex action"). Other studies that have shown referential uses of gaze include, among many, Carletta et al. (2010), Clark and Gergle (2011), and Kreysa et al. (2018). The latter has a good review of the prior literature.

Obviously gaze finds its significance from its being a social action. As Kidwell (2009) stated: "another's gaze shift constitutes a publicly available resource that offers participants opportunities to locate potentially relevant features of, and happenings in, the environment for their own attention and action" (p. 148). For example, children monitor where caregivers are looking and thus determine whether it is safe to continue misbehaving (harassing other children) or whether the caregiver(s) have noticed (seen) them doing so and will intervene. Furthermore, the harassed children may look at the caregiver but in fact accomplish different actions by virtue of what actions they perform (or stop performing) as they do so. For example, if they stop their actions the gaze counts as "noticing" the caregiver, whereas if they address them, it counts as a "search."

Gaze as a Marker of Affiliation or Aggression

Interpersonally the basic meaning of gaze is that "another [person] is attending [and their] visual channel is open" (Argyle & Cook, 1976, p. 96). Argyle and Cook (1976) reported that subjects look more at participants that they like. Bayliss et al. (2006) showed that subjects liked what other subjects looked at. This result was replicated (Tipples & Pecchinenda, 2019) but with a smaller effect size. Kleinke (1986) listed several other studies showing that gaze is interpreted as a sign of liking or attraction. However, sustained gaze may also indicate hostility (Argyle & Cook, 1976, p. 74) or dominance (think of the "staring contests" in which children engage, in which the first person to avert[5] their gaze "loses"). For a survey of the dominance gaze literature as well as some interesting studies that confirm that sustained gaze correlates with aggression or noncooperation, see Giacomantonio et al. (2018). We may speculate that the dualism of gaze, which can be both affiliative and disaffiliative, comes from its marking intimacy: if the intimacy is welcome, then the gaze is perceived as affiliative; if it is unwelcome, then it is perceived as disaffiliative.

Subjects experiencing shame, humiliation, or embarrassment are often found to avert their gaze or to look down (Argyle & Cook, 1976, pp. 77–9). Libby and Yaklevich (1973) examined personality differences of eye contact and direction of **gaze aversion**. They found the people with a "high abasement" personality (i.e., a tendency to feel guilty and accept blame) looked more to the left, regardless of whether the question is embarrassing, per se. People with a nurturing personality maintained more eye contact. Females maintained more eye contact with other females.

It has also been argued (e.g., Grossmann, 2017, p. 109 and references therein) that being looked at, or the mere presence of eyes (even just a picture), fosters cooperation.

Joint Attention

If two participants in an interaction have the same focus of attention, this is referred to as **joint attention**. Emery (2000) further distinguished between joint attention and shared attention: as we saw, the latter also requires mutual awareness of the joint attention. Pfeiffer et al. (2012) showed that "establishing shared attention takes longer and requires a larger number of gaze shifts as compared to joint attention" (p. 1). Joint and shared attention are triadic interactions (i.e., involve two participants and a third object of attention).

The research on joint attention originates from the study of child/adult interaction and particularly of how children learn to look at what the adult is looking at (Dunham & Moore, 1995; see also the papers in Moore & Dunham, 1995). Wilms et al. (2010) used eye tracking to create a computer-animated character with whom the subject could make "eye contact" and whom the subject could direct to look at an object by shifting their gaze to it. On embodied agents, see Embodied Conversational Agents section. Kreysa and M. J. Pickering (2011, p. 944) made a forceful case for the study of gaze in a naturalistic setting and specifically in a dialogic situation, as they pointed out that dialog is more primary than monologue.

Collaborative Tasks

One area of HCI involves (mostly remote) collaboration between users. Jermann et al. (2010) called their work "dual eye tracking"[6] because the gaze of both players is tracked. Theirs is one of the first examples of social eye tracking, albeit without face-to-face interaction. Nüssli (2011) was very aware of the problem of synchronization posed by the use of two devices. This can be solved in software, that is, using a third computer to log the two independent data streams, as Nüssli (2011, p. 78) showed, or by synchronizing the two data steams post hoc, by uploading them to a video annotation program such as ELAN (ELAN, 2020).

Arai et al. (2009), Jermann et al. (2010), and Nüssli (2011) used "**gaze cross-recurrence**," which measures when two participants look at the same thing. The term was introduced by Richardson and Dale (2005), in a different context; that is, they investigated how much the gaze of a speaker and of a hearer matched. Only the speaker spoke (i.e., it was a monologue). Both speakers looked at images related to the content of the speech (a TV show). Richardson and Dale (2005) found significant convergence (cross-recurrence) but with a lag of 2 seconds: in other words, the hearer's gaze matched the speaker's only about 2 seconds later. Significantly, the hearers who matched most did best on a comprehension test. In dialogic data, Arai et al. (2009, p. 1249) reported coordination between the speakers "beyond chance." Nüssli (2011) found that the coupling was related to the degree of performance (i.e., the higher the coupling, the better the performance on the task (p. 115). In Nüssli (2011) and Jermann et al. (2010), subjects played Tetris[7] collaboratively side by side but do not look much at each other. Jermann et al. found that experts and novices have different gaze patterns: whereas experts look more to the sides, the novices focus more on the center where their own falling piece is located and less on the shapes at the bottom of the field. The findings of the study are that when the players collaborate they look more at the other player's pieces.

Staudte and Crocker (2018) showed that gaze has a facilitative effect on conversation, both facilitating disambiguation, since we know that speakers look at the referent they are about to name, and providing feedback to the speaker that the hearer is following/understanding, since we know that hearers produce anticipatory glances. In other words, the speaker monitors for the hearer to match or precede the shift in gaze to the upcoming referent. Staudte and Crocker (2018) also addressed the question of whether the deictic function of gaze is automatic/reflexive in nature or whether it is an intentional/volitional strategy. They described a set of experiments that concluded that gaze is more automatic than intentional. This goes against the results by Kopp and Bergmann (2013) for synchrony (see later in the text).

Other studies have shown that collaboration is facilitated by providing the users with access to the location of the gaze of the other user. For example, Qvarfordt et al. (2005) used a shared map on which the location of the gaze of a participant, the tourist, was projected. The availability of that information was shown to facilitate the "consultant's" job of directing them. Brennan et al. (2012) showed that two participants communicating remotely but with access to their partner's gaze performed better. See also the relevant section in Amati and Brennan's useful review (2018). Research in this area continues; see for example an application to teaching writing, in Kütt et al. (2019).

Mutual Gaze

Some have distinguished between mutual gaze (in which two people look at each other's faces) and eye contact (in which two people look at each other's eyes) but, as Kleinke (1986, p. 79) noted, the distinction is probably impractical.[8] An interesting finding is that speakers overestimate the amount of eye contact that actually occurs (Rogers et al., 2018) and they went so far as to call it the "eye contact illusion" (Rogers et al., 2019). Indeed, Rogers et al. (2018) suggested replacing the expression "eye contact" with "mutual face gaze." Hessels et al. (2017) argue that there is a tendency to fixate in the eye region in face-to-face interactions. However, the situation is more complex: Rogers et al. (2018) reviewed various studies that conclude that there is significant individual variation when looking at faces, with some individual preference for the eyes or the mouth, as well as the possibility of distributing one's fixations more or less evenly between them. They call this the eye-mouth continuum. The literature also confirms that these preferences are stable (i.e., they persist in time; if retested at a later occasion, participants show the same preference) and they carry over from the laboratory experiments (usually performed by looking at images of faces in screen) into the real world.

Mutual gaze obviously indicates mutual attention. As Argyle and Cook (1976, p. 85) put it, "in many situations, 'catching someone's eye' makes it almost obligatory to start an interaction." This may be affiliative, for example, in flirting, or aggressive/dominant (i.e., a threat or a violation of social norms; pp. 92–3). Mutual gaze, or eye contact, is said to be one of the most intimate gestures. Studies have shown that eye contact increases with intimacy. However, it is mediated by personal social factors (Argyle & Cook, p. 65). Smiling, embarrassment, and other factors may also affect eye contact and intimacy. Pönkänen and Hietanen (2012) showed that eye contact causes greater emotional arousal than averted gaze and that smiling intensifies the effect.

It should be noted that Argyle and Cook (1976) also reported significant cross-cultural differences in patterns of mutual gaze; for example, while in Western societies there is a restriction of extensive mutual gaze (children have to be taught not to "stare"), among the Mende (Sierra Leone) those who avert their gaze are regarded with suspicion. Eye contact correlates with "many interpersonal factors, such as attractiveness, authority, and perceived sincerity (Argyle & Cook, 1976)." See also Richardson and Spivey (2004, pp. 14–15).

Synchrony

Speakers in an interaction tend to match the behavior of their interlocutors, not just in the sense that if someone is using, say, profanity they will do so as well, but in a much subtler sense that they will match their prosodic choices, their gestures, and bodily stance. As M. J. Pickering and Garrod put it: "in dialogue, the linguistic representations employed by the interlocutors become aligned at many levels" (2004, p. 169). This phenomenon is referred to as **synchrony**.

Dale and Spivey (2018) suggested treating the dyad as an "integrated system" to account for the "feedback loop" or bidirectional flow of information between the participants. In other words, one speaker's (re)action affects the other speaker's action and so on. For example, prosodic choices play a significant role in aligning conversational turns, not only in terms of timing but also prosodically (Szczepek Reed, 2006; Pickering & Baker, 2012).

Eye contact is believed to be the basis of joint attention (Farroni et al., 2002), and it has been suggested that eye contact is the cause of the synchrony found between interactants (Saito et al., 2010; Kinreich et al., 2017). Synchrony is associated with theory of mind, which has often been associated with mutual gaze (Baimel et al., 2015; Grossmann, 2017). Dikker et al. (2017) showed that eye contact before the start of the lecture facilitates synchrony in a classroom environment. A lot has been written about different approaches to synchrony (also under different terminology, such as lexical entrainment and alignment; see a review in Oben and Brône, 2016). One of the big issues has been whether synchrony is automatic (i.e., driven by perceptual/processing causes) or a deliberate choice made by the speaker(s) (i.e., strategic). Kopp and Bergmann (2013) showed that both mechanisms may be active, possibly concurrently. Here we use synchrony as an umbrella term for any instance of matching behavior (on any level) between speaker and hearer.

Gullberg and Holmqvist (1999, 2006) showed that hearers pay relatively little attention to gestures (less than 10 percent of gazes), focusing instead on the face of the speaker, with three exceptions: (1) gestures in the peripheral area of vision, (2) gestures that the speakers had gazed at (the reader will recall the deictic function of gaze; see Functions of Gaze section), and (3) gestures with "holds" (i.e., a pause within the gesture). Gullberg and Kita (2009) found that the position of the gesture is irrelevant but the function of gaze and the hold is significant. Oben and Brône (2015) also showed that when listeners gaze at speakers' gestures they are more likely to reproduce the gestures in their turn. Brône and Oben (2018, p. 5) also reviewed literature showing that users behave the same way toward embodied conversational agents (see Embodied Conversational Agents section), copying their hand gestures and intonation if focalized by gaze.

Mutual gaze also plays a significant part in coordinating conversations. Bavelas and Chovil (2000) and Bavelas et al. (2002) claimed that, generally speaking, a hearer will look more at the speaker than the other way around. When the speaker looks at the hearer back channeling ensues ("mm-mm," "uhuh," "Yes!" nodding, etc.). See Richardson and Spivey (2004) for more discussion. There are several overviews of mutual gaze (eye contact): Senju and Johnson (2009); Jongerius et al. (2020); Hessels et al. (2020).

Gaze Aversion

Gaze aversion can be defined as looking away from mutual gaze or from another object of attention. Gaze aversion was long believed to be caused by embarrassment or to be tied to hemispheric control: Argyle and Cook (1976, p. 23) reviewed the early evidence that questions cause gaze aversion and that the type of question (verbal, left hemisphere vs. spatial, right hemisphere) influences the direction of the shift, downward and right in the former and upward and left in the latter. However, further research showed that the primary causes for gaze aversion lie elsewhere, and primarily in the management of cognitive load, turn taking, and affective factors. In the following section we will review these areas.

Improvement of Cognitive Performance

There is compelling evidence that the frequency of gaze aversion is related to the difficulty of cognitive processing (Glenberg et al., 1998; Doherty-Sneddon et al., 2002; Doherty-Sneddon & Phelps, 2005).

This has led to the hypothesis that gaze aversion reduces visual input during high cognitive load, thus lowering the overall cognitive load, and therefore improves cognitive performance. It should be noted that in gaze aversion speakers do not avert the gaze just from the face but also from other distracting stimuli. In some cases they simply close their eyes.

Management of Turn Taking

An important function of gaze aversion is the management of turn taking. That is, when the speaker is *not* looking at the hearer they convey the message that they intend to hold on to the floor. Conversely, when they do look at the hearer at the end of a turn, they convey the idea that the hearer should, or may, talk. Gaze can also be used to select the next speaker in a conversation with more than two participants. We will examine this in much more detail in Conversational Organization section.

Affect Management

Kendon (1967), in a study based on only one subject, claimed that gaze aversion was used to modulate intimacy. Since mutual gaze is a correlate of intimacy, he postulated that there may be a threshold of acceptable intimacy or emotional involvement. When that threshold is reached, participants can lower the level of emotional involvement by averting their gaze. Kendon found an inverse correlation between intensity of smiling and duration of eye contact (i.e., the more smiling, the less eye contact). Gironzetti et al. (2016a,b) used a dual eye-tracking paradigm to analyze the gaze of two interlocutors with a focus on what happens during the episodes of humor, in particular on the direction of the gaze to the mouth and eyes; see also Gironzetti (2017). One of the findings in these studies is that increased and synchronous intensity of smiling correlates with the production/perception of humor.

Stifter and Moyer (1991) found the same pattern of arousal regulation postulated by Kendon, in infants. Bavelas et al. (1986) have shown that "motor mimicry" (e.g., wincing when someone *else* is hurt) is affected by eye contact. As Bavelas et al. (1986) and Bavelas (2007) pointed out, this supports a "micro-social" view of interactions. Stanley and Martin (1968) showed that gaze aversion occurs more frequently when the subjects are anxious and thus postulated that gaze aversion may serve an anxiety reduction function. Doherty-Sneddon and Phelps (2005) argued that gaze aversion also serves to manage social difficulty but its primary function is to manage cognitive load.

Gaze Aversion in Deception and Irony

The idea that gaze aversion, along with nervousness, is a cue to deception is unsupported by evidence; see the recent overview by Vrij and Fisher (2020)

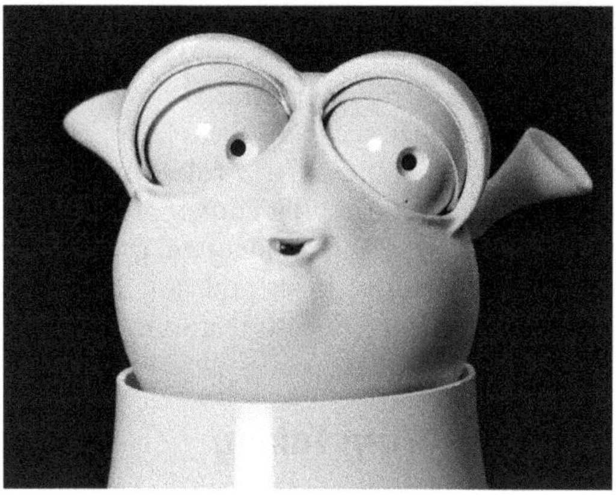

FIGURE 7.1 *The Reeti robot's averted gaze while producing an ironical turn. Reproduced with permission of Hannes Ritschel.*

and references therein. Interestingly, many law enforcement personnel are taught that gaze aversion is a clue to deception. However, there is evidence that when they are being ironical, speakers avert their gaze (Williams et al., 2009, p. 565). Gironzetti (2017) found that when they are ironical, speakers look significantly more at the mouth (rather than the eyes). Remarkably, there is now a robotic model that implements gaze aversion and other prosodic cues and facial expressions to produce multimodal ironical utterances (Ritschel et al., 2019). The project uses a Reeti[9] robot (Figure 7.1), which can be programmed to display facial expressions. The authors reported a successful implementation, with the human participants in the study finding the robot funny or ironical and preferring it to the unfunny robot. They cautioned that the cuteness of the robot may have primed the positive evaluation, however.[10]

Conversational Organization

Gaze is a significant factor in conversational organization. Obviously it is impossible to deal with the immense amount of work produced in conversation and discourse analysis. Here we limit ourselves to the role of gaze in conversation.

Engagement in the Conversation

As we have seen, there have been many claims that listeners look at the speaker more than the speaker looks at the listeners. Generally speaking, there are many reasons this may be true, if for no other reason than seeing the lip movements of the speaker facilitates listening. However, there have been claims, reviewed by Rossano (2013, p. 312), that this is a cultural issue and that, for example, African-Americans show the opposite pattern. Goodwin (1981) suggested that speakers actively solicit the hearer's gaze, for example, by pausing; in fact, he states a rule: "A speaker should obtain the gaze of his[/her] recipient during the course of a turn at talk" (p. 57). Rossano (2013) found that the requirement of gazing at the speaker is present only in extended narrative turns and not in interactive dialog (p. 314). This brings us to turn taking in conversation.

Turn Taking

A somewhat counterintuitive finding of research in turn taking is that speakers start planning their responses *before* the end of the previous turn,[11] and then they monitor the current turn for signals of its impending completion (Levinson & Torreira, 2015). Needless to say, gaze is not the only cue used by the speakers to coordinate turn taking. Intonation, pauses, semantics, and syntax are also used. The bulk of the literature on these signals is concerned with syntactic (Sacks et al., 1974) and prosodic signals (Pickering, 2018).

Turn taking was one of the topics that originally inspired conversation analysis. Sacks et al. (1974) noted that conversations minimize gaps and overlaps and are performed at "transition-relevant points" (TRPs)[12] using a "current speaker selects next" rule. These ideas have been at the center of much discussion in the context of gaze. The Sacks et al. (1974) model is based on the units that constitute turns. When a hearer identifies a turn-constitutive unit, they can anticipate that conclusion of the turn and hence the presence of a TRP (i.e., a place where turn taking is possible, without disruption). The units in this model are primarily syntactic: sentence, clause, phrase, and word.

Kendon (1967, p. 36) claimed that speakers look away at the beginning of extended turns and look at the hearer at the end, thus signaling that they are ready to yield the floor. There have been claims (e.g., Rossano, 2013, p. 315) that the idea of gaze-based speaker selection is not confirmed by additional studies. However, Kleinke (1986), Jokinen et al. (2013), Ho et al. (2015), and Auer (2018, 2021) reviewed a number of studies that confirm Kendon's thesis overall. Auer's own (2021) social eye-tracking study comes to the conclusion that "gaze is an efficient and ubiquitous way of selecting next speakers, co-occurring with other techniques or turn-allocation or alone" (p. 20).

Duncan (1972, 1974) proposed that speakers signal their intention to either relinquish or keep the floor using "signals." The signal allows the hearer ("auditor") to take the turn but does not require them to do so. Duncan and Niederehe (1974) are focused on the claiming of turns.

Among the cues that display the signal they mentioned a shift away of the direction of the head (which leads to gaze aversion), audible breath intake, gesticulation, and loudness (higher volume). Duncan and Niederehe (1974) noted that shifting the head away is compatible with Kendon's (1967) gaze aversion finding. Duncan (1974) also mentioned the turning of the head, among other signals, to claim the turns and, conversely, turning to the hearer to yield it (1974, p. 167). Duncan further noted that there is a strong correlation between number of cues (signals) and the occurrence of a turn taking so that the model is probabilistic but discrete (yes/no) (see Thórisson, 2002, p. 189, for the significance of this choice, which speeds up computation).

Duncan's approach has been criticized because turn final cues do not leave enough time for the proper timing of turn taking and because the idea of context-free signals has been rejected, for example, by conversation analysis (Sacks et al., 1974). Indeed, Duncan's "signal" approach has now been largely superseded by the conversation analytical approach but as Levinson and Torreira (2015, p. 4) pointed out it had the merit of highlighting visual cues (including gaze) and the multimodal convergence of cues, which yield a probabilistic model. However, turns tend to end at the same time as activities, which would explain the co-occurrence recorded in previous studies (e.g., Kendon, Duncan, etc.). Despite the criticisms, Duncan's work is still used, for example, in robotics (e.g., Skantze et al., 2014).

Rossano (2013) suggested that gaze is connected more with the end of activities than turns. He sees gaze as "implementing" social action depending on contextual factors (2013, p. 322). For example, the work of Kidwell (2009, 2014, 2015) has shown that gaze aversion changes meaning depending on the context, for example, whether it happens in a conversation or in a police questioning, where it is seen as disrespectful, presumably because one should give one's full attention to a police officer who is questioning them.

Thórisson (2002) presented a multimodal model of turn taking that incorporates syntactic, semantic (the content of the utterances), prosodic, gestural, and gaze information. His work is within the embodied agent paradigm (see Embodied Conversational Agents section), and so the model is ultimately computational but is nonetheless relevant to our discussion, because it is one of the broader overviews of an obviously complex, nonlinear phenomenon. The starting point is that, since there is so little time to plan and execute turn taking, most turn taking occurs within less than 100 ms of the end of the previous turn, and as 100 ms is the lowest threshold for reaction time, turn taking must be governed by "anticipatory perceptual processing" (p. 181). Some of the features that are perceived include the signals and cues described in the literature. Not only must the linguistic system be anticipatory but it must also operate in parallel, since on the one hand it is perceiving the speech stream and on the other hand it is planning what to say and when to take the turn (p. 182; see also Corps et al., 2018;

Barthel & Levinson, 2020). On prediction in linguistic processing in general see Huettig (2015).

Thórisson's (2002) model consists of three "layers," distinguished functionally but also in terms of response times.

1. **Reactive:** this layer includes gaze following, gaze aversion when wanting to take the turn (cf. Goodwin, 1981), and directing the gaze to objects mentioned by the speaker (cf. the eye-mind hypothesis). This layer is fast and is below the threshold of consciousness, i.e., < 500ms.
2. **Process control:** conscious, deliberate processes take place in this layer, such as deciding to begin or stop speaking, interruptions, etc. It is therefore slower (between 500 and 2000 ms).
3. **Content:** this layer is concerned with the content or topic of the conversation. It is the longest set of processes that can take up to minutes.

Based on this information, the layers make decisions as to what to do next. These decisions are yes/no decisions (you either interrupt or you don't). Decisions may be canceled up to 100 ms before the action is started.

Among the criteria the system monitors to determine whether to take a turn are: silence from the other speaker (i.e., the speaker has stopped speaking), syntactic completion of the unit, intonation (rising or falling), and gaze (essentially, the would-be speaker checks that the other speaker is not looking at them; see gaze aversion, below) hesitation ("(1) Saying 'ahhh . . .', (2) gazing upwards, or (3) putting on a pensive facial expression"; Thórisson, 2002, p. 201). Other models of turn taking are reviewed in Levinson and Torreira (2015).

Holler and Kendrick (2015) discussed three-person interactions in which one of the participants is unaddressed (i.e., two of the three participants are talking to each other and the third is not part of the exchange). They showed that the third unaddressed participant is not only able to anticipate turn changes but in fact they shift their gaze toward the forthcoming speaker before the end of the previous turn, presumably, as Holler and Kendrick argue, to optimize the amount of information available from *both* speakers. Zima et al. (2019) considered gaze aversion in a similar setting. They found that in cases of overlaps, the first person to avert their gaze "wins" the turn in 80 percent of the cases. The same pattern, with a slightly lower percentage, holds for co-starts: the "prevailing speaker" averts their gaze. Auer (2018) also considered three-person interactions and found that gaze is used to select the addressee (i.e., to make it clear to whom the speaker is directing their remarks) and at the end of a "turn-constructional unit" (i.e., any stretch of speech up to a TRP) selects the next speaker; see also Auer (2021).

Embodied Conversational Agents

Within HCI there has long been a trend toward anthropomorphism and embodiment, that is, interfaces that resemble humans, ranging from simple vaguely anthropomorphic characters that provide a front end to an expert system (e.g., Microsoft's infamous paper clip assistant "Clippy") to realistic CGI animation (e.g., Saya; Figure 7.2). The ultimate in embodiment is the use of an actual robot that serves as the interface to a system either self-contained (the computer is in the robot), or that the robot is controlled by.

The reason for using embodied interfaces/agents is obviously that they are more relatable than a "Help" button and presumably people will interact more with an interface that they can relate to. There have been many such interfaces, so we will limit ourselves to a few cases that show greater awareness of eye movements. For example, Greta is an embodied interactive conversational agent that has been programmed to have some emotional and communicative behaviors (Poggi et al., 2005). Pelachaud (2015) described her as follows:

> [Greta] can display complex emotions such as masking one expression of emotions by another ones [*sic*], its relationship towards its interlocutors, specific social signals such as smile and laughter. In an interaction, the agent can be a speaker or a listener. It can exhibit backchannels, mimic on the fly its interlocutor's behaviors.

In particular, Greta has an underlying model of gaze and emotional/affective expressions (Poggi et al., 2000; Poggi & Pelachaud, 2002; De Rosis et al., 2003).

We have already mentioned Thórisson's (2002) embodied agent that effectively models turn taking using a range of cues that include gaze.

FIGURE 7.2 *Saya, a realistic CGI character.* Source: *TELYUKA, reproduced with permission.*

Generally speaking, users show a preference for embodied agents that follow their gaze (Willemse & Wykowska, 2019) presumably because it is a more human-like behavior. The reasons for including a realistic model of gaze in a robot or embodied agent are not limited to the relatability of the agent, however. Nakano et al. (2003) showed that gaze is used to establish common ground, that is, shared background information, between a human user and a robot-looking agent on a screen (named MACK). In Gaze Aversion in Deception and Irony section, we saw a robot that uses gaze, among other cues, to signal irony. Skantze et al. (2014) showed that a robot's gaze facilitates a drawing task for humans, presumably because it provides joint attention, as well as facilitating turn taking. Staudte and Crocker (2011) also showed that sentence comprehension is facilitated when gaze information is available, presumably because the interlocutor attributes a referential intention to the act of gazing itself (as we saw in Functions of Gaze section on the deictic function of gaze). Mutlu et al. (2009) modeled the kind of three-person conversations we discussed in Turn Taking section, with a robot.

Facial emotion detection/simulation is obviously an important aspect of creating a believable embodied virtual agent and of interacting with human agents. We will deal with this aspect in Chapter 9. A review of studies about modeling gaze in embodied conversational agents can be found in Ruhland et al. (2015). See also Bailly et al. (2018).

Conclusion

Social eye tracking is a very new subfield of eye tracking, which has existed for about ten years, but it has made a remarkable set of strides toward becoming a full-fledged field. Social eye tracking emerged from the blending of the social study of gaze, ethnomethodology (mostly, conversation analysis), HCI, and with the availability of reasonably priced mobile or wearable eye trackers. While most of the studies examine dyadic conversations, in more recent work, attention has been paid to multiple subjects conversations (i.e., three and above). We differentiate between joint attention (onto a third party) and mutual gaze (the speakers gaze at each other). The overall picture that is created is one of great dynamism, with new areas emerging. It has been established that gaze facilitates interactions and collaboration. The evidence is also growing that synchrony plays a significant role in modulating interaction. Gaze aversion has emerged as a significant field, both because of its role in turn management, another important area in this field, but also for its function in cognitive load management. Finally, a distinctive aspect of this field of study is that embodied conversational agents and robotic applications abound and in some cases inform the theoretical discussion. This is presumably due to the uninterrupted heritage of scholars who have enriched HCI with eye tracking.

Self-Study Questions

The following questions pertain to the contents of this chapter. You should be able to answer them without looking up the answer in the book.

1. What is the difference between dyadic, triadic, three-party, and polyadic (multiparty) conversations?
2. What is the deictic function of gaze?
3. Why is the field of human-computer interaction (embodied agents, robotics) so relevant to social eye tracking?
4. How is joint attention different from mutual gaze?
5. What is synchrony?
6. How is gaze aversion used to manage turn taking?

Words to Know

The following is a list of words that have been introduced or discussed prominently in this chapter and that you should be able to define. You may test yourself on them, for example, by making flash cards.

social gaze	mutual gaze	gaze following	gaze aversion
eye contact	joint attention	shared attention	ethnomethodology
implicature	contextualization cue	synchrony	dyadic
polyadic	triadic	deictic	gaze cross-recurrence
	embodied agent	TRP	

Further Reading and Sources

The collection of articles in Brône and Oben (2018) is probably the best book-length synthesis on social eye tracking. Hessels (2020) also has a good review of the literature. Amati and Brennan (2018) is a good survey of several aspects of the field as is the introduction by the editors.

CHAPTER 8

Classroom Attention and Lx Teaching

In a sense, this is a continuation of the previous chapter on social eye tracking, since obviously teaching is a social activity. However, the setting and the "rules of engagement" of the situation are radically different. The teacher/student relationship is very different, in many respects, from that of two friends or acquaintances talking. Just to name an obvious pragmatic example, in a teaching situation insincere questions (i.e., questions for which the speaker *knows* the answer) are not only normal but also expected. Thus the rules of social interaction, and hence of gaze-related behavior, will presumably differ. In this chapter we will review the budding literature on classroom interaction and eye tracking, with a special focus on the detection of **mind wandering**. We will then consider the second language acquisition (SLA)[1] literature, one of the success stories of eye tracking in linguistics, and will close the chapter with a discussion of some applications of eye tracking to the classroom situation. In a quantitative sense, **multimedia learning**, learning strategies, and instructional design should be the central concerns of a review of the uses of eye tracking in education, since far more studies have been produced in these areas. However, the focus of the book is on the linguistic aspects of eye tracking and so our focus is on the social aspects of teaching and naturally on (Lx) language learning.

The uses of eye tracking within online and computer-based education are part of the broader field of attention-aware interfaces. These include, beyond the studies reviewed in the following, also interactive personal digital assistants (such as Alexa or Siri) and any software that autonomously assesses the user's plans and goals and tries to facilitate them. Another field that can be related to attention awareness is the field of personal informatics (Li et al., 2010), that is, the gathering of data about one's personal life for self-examination (think smartwatches or the Fitbit sensor). One may want to track one's attention while performing a task, to be alerted if one's attention wanders off task (e.g., Langner et al., 2020). A feature like this is already being deployed in smart cars that monitor the degree of alertness of the driver. In the case of personal informatics, one would look at one's attentional focus on the screen during a task and could set an alert if their attention wanders off, for example. A more general example: if one is waiting for an important package, notifications about its delivery may be welcome, whereas notifications of software upgrades are usually not a priority. Thus, if one is hard at work on a deadline, a good user-aware system would know autonomously to let one know that the package with the fairy house bird bath has arrived but would wait to notify us of the release of the 10th software patch to MS Word. Roda and Thomas (2006) provided a good introduction and theoretical background to these issues.

Attention in the Classroom

In this section we examine some studies that deal with the use of eye tracking in a live face-to-face classroom. One cannot but agree that "Eye-tracking plays an increasingly vital role in educational science " (Jarodzka et al., 2020, p. 2). We cannot hope to provide an exhaustive review of the various uses the methodology has been put to. To use one example, Lai et al. (2013) recorded eighty-one papers just in the period between 2000 and 2012 and concluded that "the eye-tracking method provides a promising channel for educational researchers to connect learning outcomes to cognitive processes" (Lai et al., 2013, p. 90). If anything, between 2012 and 2020 the number of studies in an educational context making use of eye tracking has increased. We will instead present some representative examples of applications that tie in with other aspects of eye tracking discussed earlier in the book. For a broad recent perspective on the uses of eye tracking in the classroom, see Jarodzka et al. (2020), which is the introduction to a special issue on eye tracking in educational practice.

Rosengrant et al. (2011) and Rosengrant et al. (2020) used eye-tracking glasses to record the gaze of students in a live lecture classroom. They divided the visual field into on task (the professor, the white board, quizzes, etc.) and off task (computer or phone, looking at the room, doodling, and mind wandering, here operationalized as no movement of the eyes for more than 2 seconds). The students' attention gaze was sampled every second (1 Hz). They found that, contrary to common belief, attention does not flag after the first 10 minutes and instead is sustained throughout the class.

Choi et al. (1995) argued that scan paths reflect a cognitive model (echoing Noton and Stark, 1971; see Chapter 4). This claim has been applied to classroom interaction by McIntyre and Foulsham (2018), using eye-tracking glasses. They recorded only the teacher's gaze. They found that teachers tend to have a personal scan path/cognitive model (i.e., teachers' path are more similar to their own than to others' scan paths). They also found a significant difference between expert teachers' scan paths and beginners. This is consistent with the finding that novice teachers scan a classroom scene differently from expert teachers (van den Bogert et al. 2014):

> when student-teachers notice an event it draws their full attention; possibly at the expense of failing to notice other relevant events. Experienced teachers attend to the event, but keep monitoring the classroom. Thus, they maintain a broad overview of the classroom. (p. 215)

Wolff et al. (2016) also found differences between expert and novice teachers. For example, expert teachers monitor more areas of the classroom. Other studies using mobile eye tracking confirmed the differences between expert and novice teacher: experts can monitor the classroom *while* giving feedback to individual students (Cortina et al., 2015). McIntyre et al. (2017) confirmed the differences in expertise but found cross-cultural differences between teachers in the UK and Hong Kong. UK teachers used their gaze more for information seeking, while Hong Kong teachers used it more for information giving.

A teacher's gaze directed at students is an immediacy factor and conveys interest and involvement. Eye contact (see Chapter 7) is a mechanism to establish joint attention. Haataja et al. (2020) addressed gaze between teacher and students. They find that both teachers and students tended to look at the face of their interlocutors when they have greater agency (e.g., when asking a question). In general, most of the time teachers looked at students' faces and vice versa, without making eye contact. Interestingly, teacher started eye contact was the least frequent category, with an almost double number of student initiated eye contact.

Tutoring

An old education adage has it that the best lessons are taught on a seesaw, the children's game quintessentially played by two people. In higher education this model is embodied in the tradition of tutorials à la Oxford-Cambridge in which a student meets individually with a tutor. Needless to say, the model is prohibitively expensive. Hence the interest is in developing **intelligent tutoring systems** that might provide the one-on-one attention and customization of teaching that individual tutoring provides. The bottleneck to the creation of an intelligent tutoring system is clearly that of knowing how the student is doing (what is known as a **student model**). In adaptive testing this is done implicitly by assessing the performance of the subject up to a given point and selecting the next item based on their performance. For example, if the subject has scored correctly up to that point, a harder question will be asked, and so on. In other words, the performance of the student on the test *is* the model of the student used by the system to calibrate the test. The GRE, a test required by many graduate programs in the United States, has been adaptive since 1993, for example.

While this approach works well with testing, in tutoring one cannot check directly the level of understanding of the learner, short of using queries that might end up distracting the learners. Hence the idea of using passive technologies, such as eye tracking, to monitor the attention level of learners. Gluck et al. (2000) proposed using eye tracking to help build the student model. For example, in a mathematical tutorial system, they monitored if the learner shifted incorrectly from one column to the other. They were then able to predict some of the errors based on the pattern of fixations. Another aspect is error messages. Over 40 percent of error messages by the tutorial system were ignored by the learners. The eye-tracking component could monitor this fact and alert the learner that the message was there (e.g., with an auditory cue). Anderson (2002) developed the idea and showed that students' performance improved when they were provided with an auditory clue to "read the message" being displayed on screen after a few seconds or were given cues based on whether they were looking in the right place for the answer. Conati and Merten (2007) investigated self-explanation, that is, the capacity of a learner to clarify concepts to themselves, by working out an explanation independently. They found that adding an eye-tracking component (i.e., where the user was looking) improved the capacity of their model to assess the capacity of the users for self-explanation, but that pupil dilation did not.

Several systems have been developed to facilitate reading/learning from text. Sibert et al. (2000) proposed a system used to improve reading that tracks eye gaze on words during reading. When the reader stays on a word above a certain threshold the system pronounces the word for them. The authors report positive results and that the users (children, some of whom had dyslexia) reported liking the experience. Sibert et al. also tracked mind wandering. Indeed, a significant part of intelligent tutoring is directed to the problem of mind wandering (see Mind Wandering and Its Detection section). Conversely, the level of excitement of the subjects can also be tracked using gaze methods (Abdessalem et al. 2019). Conati et al. (2013) surveyed work on student modeling in intelligent tutoring using eye tracking. In particular they reviewed work on using eye-tracking information to model students' learning offline (i.e., the results are used for research or to improve the system but not in real time). D'Mello et al. (2012) also contained a survey of work on intelligent tutoring, as well as a presentation of their system *Guru*. D'Mello (2019) also discussed *Guru* and *Gaze Tutor*. Lallé et al. (2019) showed that data visualization helps understanding of content. Recent approaches leverage machine learning to assess, for example, learner's confusion, by comparing scan paths (Sims & Conati, 2020). Since most of these studies focus on remediating mind wandering, they will be discussed in more detail in Mind Wandering and Its Detection section.

Mind Wandering and Its Detection

Of particular interest in the study of attention in the classroom is the phenomenon of mind wandering, which involves a decoupling (Smallwood, 2011) of overt and covert attention. Mind wandering, also known as daydreaming or zoning out, is a common experience. In fact it has been claimed that one may be mind wandering up to 50 percent of their time awake (Killingsworth & Gilbert, 2010). Other studies put the rate of mind wandering between 30 percent and 60 percent of the time (Wammes et al., 2018). Mind wandering should not be confused with distraction, in which another stimulus competes with the on-task stimulus (e.g., being distracted by the TV while studying). In mind wandering there is no external stimulus to distract the subject; the distracting thoughts are generated internally by the subject (the term "self-generated" is often used in this respect). However, mind wandering is also associated with unhappiness, craving food, and alcohol intoxication (Smallwood & Schooler, 2015, p. 493), which may not directly distract the subject but may prime distraction.[2] Mind wandering is particularly common in automatic tasks, such as driving. It is not an uncommon experience to realize that one has been driving for several minutes but has been thinking about something else entirely.

Technically, mind wandering can be defined as mental activity unrelated to the cognitive task at hand. A term often used is "task-unrelated thought." Smallwood et al. (2008) distinguished between deliberate mind wandering (tuning out) and mind wandering that occurs unbeknownst to the subject (zoning out). Mind wandering is usually investigated by "**mind probes**": the investigator stops the activity and asks the subject if they were thinking about the activity or something else. Eye tracking, fMRI, and EEG offer opportunities to investigate mind wandering directly by observing physical correlates of the brain activity of the subject.

The effects of mind wandering can be serious; for example, it has been associated with car crashes (He et al., 2011; Galéra et al., 2012). Generally speaking, however, the effect is less dramatically that of impaired cognitive performance. Smallwood et al. (2008) reported that subjects who had been mind wandering at significant stages of the reading of a Sherlock Holmes story were less likely to identify the villain. Besides impairing reading comprehension, other effects of mind wandering that have been reported are impaired recollection of a text, low alertness, poor performance on aptitude tests, and poor academic performance (see later in the text). Some positive effects of mind wandering have been reported as well: it obviously alleviates the boredom inherent in repetitive tasks, and it has been associated with creative thinking (Fox & Beaty, 2019). For a review of the effects of mind wandering see Mooneyham and Schooler (2013). The meta-analysis by

Randall et al. (2014) showed that people with fewer "cognitive resources" (i.e., expertise in the task) mind wander more frequently, and that those who mind wander more frequently perform less well on the task. Mind wandering has been linked to ADHD (Smallwood et al., 2007; Seli et al., 2015; Franklin et al., 2017) and OCD, in the case of spontaneous mind wandering (Seli et al., 2017).

These effects have led to the "decoupling" hypothesis (Foulsham et al., 2013, p. 51): "the link between word identification and eye scanning is decoupled when the mind wanders." Smallwood (2011) presented a cascading model of decoupled attention in reading, which provides a good theoretical framework to consider mind wandering. Decoupled (distracted) attention causes reduced processing of the perceptive level (features such as shape, color, position, etc.); this reduced perceptual detail causes less processing by the brain of the perception, which in turn degrades higher, semantic (task-related) processing. Because the brain has less to work with it has a greater probability of forgetting or of not comprehending (e.g., what is being read).

> As mind-wandering occurs during reading, top-down attention shifts from the task. Simultaneously, the degraded perceptual representations will prevent detailed lexical processing of a word, which will then impair the creation of the propositions that make a sentence. The absence of bottom-up information processing hinders the ability of the reader to create more complex propositional and situational model of the narrative. (Smallwood, 2011, p. 70)

With consideration to all due differences, this model applies equally well to other nonreading mind wandering. Other interpretations of mind wandering include the failure to maintain control of one's attention (McVay & Kane, 2009) and a failure to allocate cognitive resources effectively (Randall et al., 2014).

Mind Wandering in Lectures

In particular, mind wandering is an issue in asynchronous online education, where the instructor cannot check where the attention of the students is directed (Jang et al., 2020). The prevalence of mind wandering in video lectures is quite high: 40 to 45 percent (Risko et al., 2012; Spuznar et al., 2013; Kane et al., 2017). Forrin et al. (2021) also report that longer (two to six sentences) stretches of text lead to more mind wandering than reading one sentence at a time. Foulsham et al. (2013) reported that mind wandering increases with time (i.e., the longer the lecture, the more mind wandering). Generally speaking, mind wandering increases with the length of the lecture.

Kane et al. (2017) showed that mind wandering increases in the second quarter of the 52-minute video lecture they used in their study and is constant thereafter. Kim et al. (2014) have claimed that most MOOC lectures are abandoned 6–8 minutes into the lecture. In their data, 55 percent of video lectures are abandoned before the end.[3]

It has been established that there is a negative correlation between mind wandering and retention of lectures. Wammes et al. (2016) reported that mind wandering negatively affects retention but less so performance on tests[4] in live lectures. Risko et al. (2012) and Seli et al. (2016) found that self-reported motivated students engage in less mind wandering (regardless of the intentional or unintentional nature thereof), and consequently they have better retention of the lectures' content. Seli et al. (2019) found that increasing motivation reduces mind wandering (again, of both intentional and unintentional kinds). Szpunar et al. (2013) found that frequent testing (within each lecture) led to less mind wandering (although less so for high school students, Szpunar et al., 2014), better performance in the test, and less anxiety and better performance toward the final cumulative test. Wammes and Smilek (2017) reported that video lectures lead to more wind wandering than live lectures. Kane et al. (2017) reported that (mandated) note taking made no difference in mind wandering frequency; however, students who took better notes experienced less mind wandering. Conversely media multitasking (e.g., checking one's phone) increased mind wandering. Mind wandering correlated with lower scores on the final test. Hollis and Was (2016) also reported significant rates of mind wandering connected with social media use.

Wammes et al. (2018, p. 261) reported that a significant percentage of mind wandering in live lectures is deliberate (20 percent vs. 15 percent of involuntary mind wandering). This means that increasing the motivation of the students will likely reduce mind wandering. Testing interpolated in the lecture may likewise have the same effect (Szpunar et al. 2013, 2014). Increased testing led, predictably, to increased note taking. In conclusion, mind wandering is a serious problem and since a good percentage of it is deliberate, it may be possible to palliate the problem with interventions, for which eye tracking is ideally positioned.

Mind Wandering and Eye Tracking

Although, Walcher et al. (2017) did not, strictly speaking, use mind-wandering data (as the task used in the study is idea generation, which causes "staring into space") it is commonly quoted in the mind-wandering literature. Walcher et al. reported increased blinking, longer blinks, reduced number of microsaccades, increased and shorter fixations, increased number and amplitude of saccades (cf. Reichle et al. 2010 "erratic" eye movements), and increased vergence (see Chapter 4) without external stimuli. Krasich et al. (2018) examined subjects looking at images. They found higher blink rate, fewer fixations, longer duration of the fixations, and "fixation dispersion." However, most of the studies of mind wandering and eye tracking have been done using reading and lecture watching data.

Mind Wandering While Reading

Unsurprisingly mind wandering during reading has been associated with comprehension failure (Smallwood et al. 2007). An interesting effect of mind wandering on reading is that during mind wandering the normal effect of word length and frequency (see Chapter 5) is diminished. For example, Reichle et al. (2010) recorded four undergraduates reading a Jane Austen novel.[5] The mean mind wandering was 22 percent, 9 percent of which subjects were unaware of. They recorded fewer lexical effects on fixation durations and longer mean fixation duration as well as more "erratic" eye movements during mind wandering (such as fixating off the line of text; see Figure 8.1).

Uzzaman and Joordens (2011) reported similar findings using twenty-two participants and a shorter text (sixteen pages of *War and Peace*). Smilek et al. (2010) reported significantly increased blinking and decreased number of fixations during mind wandering in a reading task but no significant difference in duration of fixations. Grandchamp et al. (2014) also reported increased blinking. Conversely, Jang et al. (2020) did not find increased blinking. Other effects include slower reading time (Smallwood, 2011,

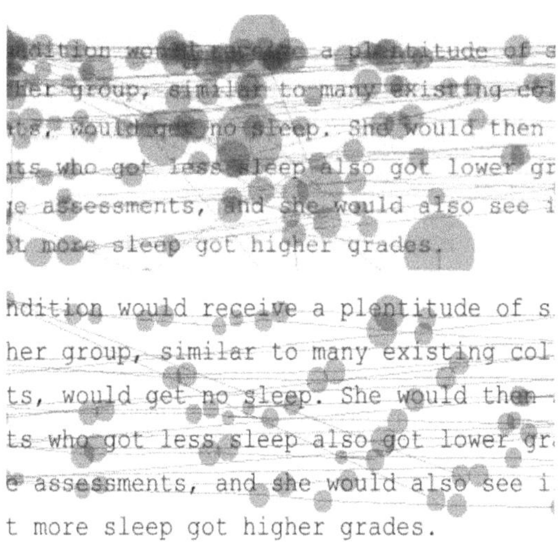

FIGURE 8.1 *Fixations while mind wandering (top) and normal reading. Sampling at 120 Hz.* Source: *D'Mello et al. (2017, p. 10), reproduced with permission of John Benjamins Publishing.*

p. 71). Hutt et al. (2016) used a number of gaze features, such as longer gaps between fixations and shorter fixations before mind wandering, captured by a low-cost eye tracker, as part of an intelligent tutoring system called *Guru Tutor* (see next section).

Mind wandering is associated with higher baseline pupil dilation. We distinguish between baseline and task-related pupil dilation. The two are inversely correlated (Gilzenrat et al., 2010); however, see Unsworth and Robinson (2018) for contrasting data. For example, suppose you are asked to read a passage from a book. The researcher measures your baseline pupil dilation before starting. As you get going, your pupil dilates, because you are on task. However, the passage is not very interesting and your attention lags: your baseline pupil dilation increases and your peak dilation decreases. Your chances of mind wandering increase as well. Similar results are also reported in Smallwood et al. (2012) and Franklin et al. (2013). Contrasting results were presented by Grandchamp et al. (2014) and Mittner et al. (2014), but they are possibly reconciled to the previous studies by Unsworth and Robinson (2018), who found that nonalert mind wandering ("mind blanking") is associated with low arousal and low pupil dilation.

The opposite of mind wandering is mindfulness. Mindfulness has been suggested as a remedy to counter mind wandering. On mindfulness as a solution to mind wandering, see Mrazek et al. (2012). Another state that counters mind wandering is **flow** (Csikszentmihalyi, 1990). On the interplay of mindfulness and flow see Sheldon et al. (2015).

Automatic Detection of Mind Wandering Using Eye Tracking Technology

The earliest implementations of eye tracking methodology in reading tutoring systems go back to Sibert et al. (2000) who used real-time eye-tracking data to monitor dwell time on a given word and provided the pronunciation if a threshold was exceeded. So the idea of using eye tracking to monitor and correct student behavior has been around for a while. D'Mello et al. (2012) implemented a tutorial system (*GazeTutor*) that uses eye-tracking information to check that the student is looking at the correct material and provides a reminder if they are not. The study showed that the agent was successful in redirecting gaze to the target areas and that the students' learning improved, as compared to the control group, who did not have the eye-tracking features in the tutor. D'Mello et al. (2017) also implemented a system that recognizes mind wandering in real time and redirects the attention of the reader. D'Mello (2019) provided a survey of GazeTutor, Guru, and other systems.

The feasibility of automatic detection of mind wandering using the ocular cues discussed earlier was established by Bixler and D'Mello (2016). Hutt et al. (2019) showed that it is feasible with low-cost eye trackers, thus making practical implementation feasible, in their *Guru Tutor* system. The system was deployed with a sample of high school students. The results show that a 40 percent accuracy can be obtained with low-cost off-the-shelf equipment: Hutt et al. used a gaming eye tracker priced at $200. The performance of the students improved in the system condition. Mills et al. (2020) described a system that tracks mind wandering during reading using eye tracking and asks the viewer to describe the situation. If the viewer fails, they are instructed to go back to the text.

Wang et al. (2006) used a head-mounted eye-tracking system to track attention, as well as other factors, and an animated character to provide feedback to the user when their attention wandered. The users were also able to use their gaze to interact with the system (e.g., choosing a subject; see Chapter 9 for other examples of this approach).

The possibility of gaze-aware instructional media raises the obvious issue of compliance. It may prove difficult for adults to accept the intrusion of privacy implicit in the technology. Wang et al. (2006) reported that some participants did not like the warning of low attention they received. Furthermore, it may be possible to defeat the technology, for example, by wearing glasses that block IR light.

Lx Acquisition

In the introduction to the special issue of *Studies in Second Language Acquisition*, which can be argued to have mainstreamed eye tracking to the field of SLA, Winke et al. (2013) stated almost a decade ago, "more and more attention is being paid to [eye tracking] by SLA researchers." That statement was correct at the time and today it is even more so. Eye tracking has become increasingly the "go-to" methodology in Lx acquisition. Godfroid et al. (2020) argue that

> since 2013, researchers in more and more subdisciplines in L2 research have taken up eye tracking. Sentence processing remains an important area of L2 research, but it is no longer the sole or dominant focus for research. SLA researchers' contemporary applications of eye-tracking technology (. . .) frequently use long and/or authentic materials in an attempt to capture the dynamics of L2 processing and learning more faithfully. As the scope of eye-tracking research grows, we see eye-tracking data being paired in novel ways with other data collection methods, including qualitative methods. Eye-tracking methodology is diversifying and breaking boundaries, its innovation fueled by different disciplinary traditions. *Within SLA, eye tracking has become a popular tool.* Godfroid et al. (2020, p. 245; our emphasis, SA/LP)

The popularity of eye tracking is not only due to the possibility of using qualitative methods but also because since early on (e.g., Roberts &

Siyanova-Chanturia, 2013) researchers realized that eye tracking afforded the use of more ecologically valid data:

> [eye tracking] allows for the study of moment-by-moment processing decisions during natural, uninterrupted comprehension, and critically, without the need to rely on participants' strategic or metalinguistic responses (. . .) This is important given that (. . .) it may be task conditions that specifically force participants to attend to the (grammatical) details of the input where learners' processing is most nativelike [. . .] or where individual differences in factors such as proficiency, cognitive capacity, and processing speed may come into play. (Roberts & Siyanova-Chanturia, 2013, p. 214)

In what follows we will examine a few of the salient areas of research, with an emphasis on the seminal articles in the field.

Prosody

Connell et al. (2018) used a visual world paradigm experimental design in which the words used as stimuli are not represented by images but rather orthographically. The study investigated whether prosodic cues to lexical information transfer across languages. The setup of the study comes from an earlier study by Cooper et al. (2002) that did not use eye tracking. As Cooper et al. (2002) explained,

> in Dutch, the words *octopus* and *oktober* ('October') each begin with the same first four phonemes, but in *OCtopus* stress falls on the first syllable while in *okTOber* it falls on the second (henceforth, upper case signals stress placement). This means that the first syllable of octopus will be somewhat longer, may be somewhat louder, and will exhibit greater pitch movement than the first syllable of *oktober* (with the reverse being the case for the second syllables). If listeners can use these aspects of suprasegmental structure to distinguish one word from another, then they could tell the difference between *octopus* and *oktober before*[6] the arrival of distinctive segmental information (the fifth phoneme). (p. 208)

In other words, the stress information will facilitate the identification of the word, which will result in faster recognition/activation. However, if the stress does *not* match in the two words, identification will be slower, despite the segmental information being identical.[7] Indeed, Cooper et al. (2002) found that ["admi-] activates more "admiral" (which has primary stress on the first syllable, indicated by the " " " mark) than "admiration" (which has the primary stress on the third syllable).

Connell et al. (2018) applied this setup to Lx speakers of English[8] and presented the stimuli as a four words: the target (e.g., admiral), the competitor (e.g., admiration), and two distractors, in a 2 × 2 grid pattern. The grid itself was not displayed. The four-word grid was displayed for 4 seconds so that the participants would activate their phonological representations. Then a fixation cross replaced the four words for 500 ms. The cross would then be replaced by the same four words, in the same position. The audio stimulus (the word) would be played at the same time. Thus Connell et al. could measure precisely toward which of the four words the gaze would be directed. The participants would click on the stimulus word, as instructed, and this would conclude the item. The experiment consisted of more than sixty items.

It is a well-established finding that speakers of an L1 that does not have lexical stress (more precisely, where lexical stress is not contrastive) such as French, for example, in which most words are stressed on the last syllable, are not sensitive to this feature. In other words, languages without contrastive lexical stress do not use it to recognize words unlike languages that do show distinctive stress, such as English, for example, in which *còmpact* and *compàct* are different words. Connell et al. (2018) contrasted L1 English speakers with Lx speakers of Mandarin and Korean. Their results can be represented in Table 8.1.

English speakers are able to use *both* suprasegmental clues alone or segmental and suprasegmental clues in combination (as shown in Cooper et al. 2002), whereas Mandarin speakers use only suprasegmental cues. Korean speakers only use segmental cues if they occur together with suprasegmental cues. These differences are attributed to the characteristics of their respective languages, which transfer into Lx. In English, lexical stress is marked by (a) the reduction of most unstressed vowels (segmental cue) and (b) higher pitch (main formant, F0) and intensity (volume) and longer duration (suprasegmental cues). Thus Mandarin-Chinese speakers are able to use suprasegmental cues because of the presence of lexical tones in Chinese, which transfer on the pitch variation of English stress. Korean speakers have a schwa sound in the phonological inventory of their language and thus they can be expected to be able to transfer that distinction into the recognition of English word stress (Connell et al., 2018).

TABLE 8.1 Connell et al.'s (2018) results

English	Mandarin	Korean
suprasegmental cues segmental + suprasegmental	suprasegmental	segmental + suprasegmental
	lexical tones	reduced vowels

Vocabulary

One of the areas of most significant interest is incidental lexical acquisition during reading (i.e., the focus of the activity is not the teaching of the lexical items). Studies have considered texts ranging in size from the sentence to the novel. For example, let us consider Whitford and Titone's (2012) study of paragraph reading. As one would expect, they found that participants had shorter fixations in L1 and read faster. They also found that L2 word frequency effects (see Chapter 5) in L2 were greater than in L1 for bilinguals and that there is a correlation between amount of exposure to L2 and size of the word frequency effect. They interpreted these results as supporting implicit learning theory (insofar as exposure correlates with proficiency). Similar results had been found for words in isolation and in sentences (see Chapter 5) and for bilinguals (e.g., Duyck et al., 2008). However, Duyck et al. (2008) used words in isolation, in their recognition task. Whitford and Titone (2012) showed that these results also apply to words contextualized in a paragraph. On lexical processing in bilinguals see also Santesteban and Schwieter (2020) and Hayakawa and Marian (2020).

An area that has emerged only very recently is reading-while-listening: with the widespread availability of audiobooks in recent years it has become a practice to ask students to read a book, for example, while listening to the audiobook. Conklin et al. (2020) reviewed several studies that show various benefits including increased fluency, increased vocabulary learning, and an overall better experience than just reading. In an eye-tracking study they investigated the performance of native speakers and non-native speakers reading two short (1,500 words) stories, varying the condition of listening/nonlistening to the audio while they read. They found that reading while listening to the audio slowed down L1 readers but had no impact on L2 readers: they were generally slower than L1 readers but were not further slowed down by the audio. In terms of fixation durations, the mean fixation of L1 readers was shorter than those of L2 readers, in the silent reading case. However, in the reading-while-listening, they were not significantly different (i.e., the NS advantage disappears) and crucially they matched the reading-aloud figures established in the literature. Thus Conklin et al. concluded that reading aloud and reading-while-listening causes the reader to monitor their reading speed, so as not to get ahead of what they are listening/saying (reading is faster than speaking in L1). Since L2 readers are slower than L1 readers to begin with, reading out loud or while listening has the effect of eliminating the difference. Finally, in terms of alignment L2 readers' fixations were more aligned (on the right word) than L1 speakers, showing that they "are making more use of the visual text to aid listening than L1 comprehenders and/or that the speed of the audio was better matched to the L2 readers' reading rate" (Conklin et al., 2020, p. 273).

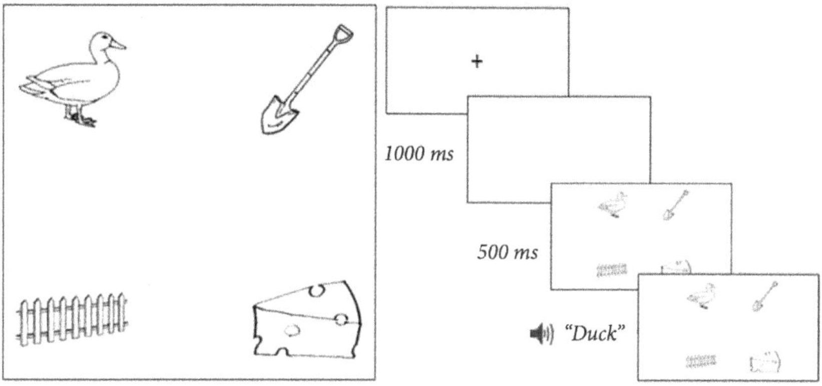

FIGURE 8.2 *The visual display in Shook and Marian (2019). Permission granted by John Benjamins Press.*

As seen in the survey in Hayakawa and Marian (2020), lexical activation in bilinguals activates all languages that the individual has access to in parallel. For example, in Shook and Marian (2019), the Spanish-English bilingual participants and English monolinguals looked at the board in Figure 8.2 and heard the English word "duck" (no Spanish was used anywhere in the experiment). However, bilinguals looked more at the shovel competitor than English monolinguals.

This may seem baffling, until one recalls that in Spanish "duck" is *pato* and shovel is *pala*. In other words, the target and the competitor *pala* overlap phonologically at the beginning of the word ([pa]) in Spanish. The reader will recall that when a word is activated its cohort is activated as well. Hence this is evidence that activation happens for the Spanish lexical cohort as well. This phenomenon has been called **covert lexical activation.** An area in which parallel lexical activation is particularly evident is idioms and formulaic language. As we saw in Chapter 5, fixed and idiomatic expressions have a high degree of predictability and may be stored lexically as a single unit and hence are processed faster. Non-native speakers process idioms somewhat differently (Carrol & Conklin, 2017). For example the literal (nonfigurative) meaning of an idiom is going to be more salient (and thus get processed first; Cieślicka et al., 2014).

An area that has attracted some interest is eye-tracking studies of dictionary use, both by L1 and Lx speakers. Lew et al. (2013) studied users of bilingual dictionaries. They found that most errors (picking the wrong sense of a word) came from selecting the first entry, a previously well-documented phenomenon. They also found that participants read dictionary entries roughly at the same speed as any other text. A typical search would briefly scan the headwords discriminating the various meanings, and once the users had found what they were looking for, they would engage in a more extended perusal of the entry for the selected meaning. In particular, the study shows that the headwords discriminating the various meanings work well, and the users pay attention to them.

However, the rise of electronic dictionaries, and of tools like Google Translate, has significantly changed the playing field. First, consulting an online dictionary is faster and hence it leads to increased use (Collins, 2017). Also, it seems that information retrieval on a web page is different than in print (Simonsen, 2011); for example, users spent considerable time looking at the window in which they input the word to be searched. Once the word has been retrieved, they performed a completely different searching process to identify the sense they needed within the dictionary entry. Tono (2011) showed that when users look up a word in an electronic dictionary the complexity of the definition is the main deterrent to the reader's reading till the end of the definition and hence if the information sought by the learner is at the end of a complex entry, it will not be retrieved. Furthermore, he shows that the syntactic information provided in the traditional notation used in linguistics (such as SVO, i.e., subject, verb, object) is not used by the readers, whereas the alternative notation "make A B"[9] is fixated on. See also Töpel (2014) survey of studies of online dictionaries and Lew and De Schryver (2014) and Lew (2015).

Syntax

There are numerous studies that focus on syntax in an SLA setting. We will use Winke (2013) as an example. The subject of the paper is the effectiveness of input enhancement proposed by Sharwood Smith (1993). Sharwood Smith started from the point that learners are metaphorically sitting in a "language bath," that is, a barrage of language in Lx, which they could not possibly assimilate in its entirety. This is particularly true today, thirty years after the paper was first written, when it is possible to access on the internet literally more material in any language than one could possibly ever hope to read/watch/listen to in a lifetime. So, the question becomes, how does the learner choose what to focus on? Sharwood Smith (1993) saw input enhancement (or "consciousness raising") as an attempt to "direct the learner's attention in particular directions" (p. 176). It is easy to see that eye tracking, as per the eye-mind theory, is ideally situated to contribute empirically to the theory of input enhancement/selection. Input enhancement may be positive (such as bolding or highlighting certain forms in the text, as we do in this book!) or negative (error correction). It can be explicit (e.g., drawing attention to a given form metalinguistically) or implicit (e.g., using a different color for characters in an otherwise monochromatic text) and elaborate or not, depending on the length of the process, which can range from the few milliseconds needed to read an italicized word/morpheme to minutes or hours spent doing an activity in class, for example.

The research on the effectiveness of input enhancement shows mixed results. Winke (2013) provided a long list of factors that may affect how well enhancement works, including the type of enhancement, the readiness of the learner, and the comprehensibility of the text, to name a few examples. This study is interesting because it shows forcefully the usefulness of using eye tracking to replicate prior studies. As Winke noted, previous studies could only measure the effects of enhancement indirectly (e.g., by measuring production or learning or through think-aloud and recall protocols), whereas eye tracking allows the researcher to view whether the attention of the reader has been captured or not, directly.

To do so, Winke replicated a prior study by Lee (2007) in which learners were given two versions of a text with numerous examples of passive voice. In one version, the passives were highlighted. After they had read the passage they were tested on the passive, using a correction task in which they were presented with incorrect passives and had to recognize them. The students who had received the enhanced version outperformed the students who had read the normal version of the text; however, the students who had received the enhanced version also underperformed on a measure of comprehension of the text.

Winke (2013) enhanced the passive forms of her passage by underlining them and displaying them in red characters. Winke then measured, using an eye tracker, total fixation time, first-pass reading time, rereading time, and number of visits to the AOI for each instance of passive. She found significant increases for total fixation time and rereading time. No other measures yielded significant differences. However, Winke noticed a decline of the enhancement in each slide: the text was divided in four slides, for practical reasons of visibility on screen. Each new slide caused an increase in fixation time on the enhanced forms but the effect diminished, as the reading of the slide advanced. So, in conclusion, the eye-tracking study showed that enhancement does succeed, at least initially, in drawing the learners' attention to the enhanced forms. The students' performance on the comprehension tests did not show any significant differences. Thus, Lee's finding that there was a trade-off between attention and comprehension was not replicated, but also no positive effect of enhancement on learning was found. Winke noted that one possible explanation is that regressions and longer fixations may be caused not only by increased attention but also by cognitive processing problems.

Prediction

As we saw in Chapter 6, speakers do not passively wait for information to be provided, but they anticipate or predict what is about to come. The question in SLA is whether L1 and Lx predictions differ. We know that there are different kinds of predictivity in language (Huettig, 2015).

At a very basic level, the **noticing hypothesis** (Schmidt, 1990) presupposes some level of prediction: in order for the learner to notice that there is a mismatch (a "gap") between what they have said and what a native speaker would have said, the learner has to have produced a prediction of what the native speaker was about to say (Godfroid, 2020, p. 104). The noticing hypothesis, in its simplest formulation, is that "noticing is necessary for SLA" (Schmidt, 2012). The significance of the noticing hypothesis is that it addresses the central issue of the long-standing debate between explicit and implicit learning, falling squarely on the explicit side. In other words, the issue at stake is whether mere perception (no consciousness) is enough for learning or whether some awareness (noticing) is necessary.

It should be noted that attention is equivalent to noticing (noticing is the "subjective" equivalent of attention; Schmidt, 2012), and thus the subject lends itself to eye tracking. So, Godfroid et al. (2010) set out to measure the degree of noticing that learners have of some words in a silent reading that includes unknown words. In fact, the words were pseudowords, to ensure that they were absolutely unknown and had a null frequency, since obviously they have never been encountered previously, while maintaining a plausible phonotactic structure; that is, they would pass as possible words (e.g., *paniplines*). By recording the fixation times and the scan paths of the readers, Godfroid et al. were able to identify noticings, which display longer dwell times and more regressions: "The reader was thus cognitively engaging with the target word, which we interpret as a noticing event" (Godfroid et al. 2010, p. 183). The study showed that mere perception (i.e., no noticing) was enough to recognize the word but not to learn it. In a follow-up study, Godfroid et al. (2013) showed a correlation between noticing and learning a new word. Of particular interest is the discussion of the advantages of eye tracking over prior methods to track noticing: asking the readers to underline significant words, verbal reports, and written note taking.

A more complex kind of prediction is related to the semantic prediction that speakers make while processing utterances. The underlying idea is that Lx learners because of lack of knowledge or of processing load would be unable to make as many predictions. This is known as the RAGE[10] hypothesis. The rage hypothesis is predicated on the basis of the work of Kaan[11] and others (Kaan et al. 2010; Kaan, 2014). In Kaan's version, the RAGE hypothesis can be presented in a strong version (Lx speakers make no predictions, based on syntax; Kaan et al. 2010) and in a weak version (Lx speakers make fewer predictions; Kaan, 2014). In Kaan's view, Lx speakers build the same syntactic structures as those of L1 speakers.

Reduced predictions for Lx speakers have been found in prosody (Perdomo & Kaan, 2019), syntax (Kaan et al., 2010; Kaan, 2014), and in semantics. Godfroid (2020) provided a review of the evidence of prediction based on both semantic cues (see Chapter 6), in which the evidence is that Lx speakers use the same mechanisms than L1 speakers do, and morphosyntactic cues: L1 speakers of languages with grammatical gender, such as Spanish or Italian, make predictions based on the gender of the article, whereas Lx speakers do not, or start showing L1-like behavior only when very advanced. There is also evidence that gender prediction transfers from L1 into Lx; in other words, if the two languages have the same grammatical markers for gender the Lx speaker performs as well as the L1 speakers.

However, predictive reasoning is a load on cognitive processing and so Godfroid concluded that predictions may be easier when the cognitive load is reduced (2020, p. 107). Thus predictions seem to display a direct correlation to proficiency: the higher the proficiency the more nativelike the performance in predictive tasks.

Reference

The Tanenhaus et al. (1995) paradigm of referential syntactic ambiguity (see Chapter 6) has also been applied to Lx speakers, highlighting the by now familiar effect that Lx speakers encountered more difficulty in recovering from the first erroneous interpretation. Pozzan and Trueswell (2016) contrasted the capacity of native speakers, children, and adult non-native speakers to recover from a garden-path sentence, such as "The horse raced past the barn fell" (see Chapter 5). Children have been found to be essentially unable to recover from a garden path, due to "immature cognitive control and executive function (. . .) skills" (Pozzan Trueswell, 2016, p. 2). Executive functions are attention control, cognitive inhibition (tuning out irrelevant stimuli), inhibitory control (i.e., the capacity to inhibit one's reactions when appropriate or useful), cognitive flexibility, and problem solving, for example. The Stroop test (see Chapter 6) is used to assess executive function. The idea is that executive functions are used "when the processing systems needs to abandon an initially preferred analysis in favor of a dispreferred one" (p. 2) and thus adults, in either the L1 or Lx condition, should perform unlike children, who have not fully developed their executive function. Ratiu and Azuma (2015) argued that there is no working memory advantage for bilinguals (recall, that working memory is part of executive control).

Pozzan and Trueswell (2016) contrasted two hypotheses of garden-path sentence recovery in Lx learners: (1) the cognitive immaturity hypothesis, which predicts that adults Lx learner should not experience trouble recovering from a garden-path sentence, since their executive functions are fully developed; and (2) a cognitive load hypothesis, which claims that difficulties in processing garden-path sentences come from their cognitive load (garden-path sentences are harder to process), which would lead to the prediction that Lx learners, especially the less proficient ones, should experience some trouble recovering from a syntactically ambiguous garden-path sentence.

Like in the Tanenhaus et al. (1995) experiment, Pozzan and Trueswell (2016) gave their participants sentences that were initially ambiguous until the ambiguity was resolved by the final part of the sentence (e.g., *Put the frog on the napkin into the box*).[12] The subject were Italian speakers who were learning English as Lx and English L1 native speakers. The participants' gaze was tracked while they performed actions on a computer screen using the mouse, such as moving a frog from the napkin into the box. The results show that Lx speakers made more errors in the ambiguous cases in which there was only one frog. Significantly, the percentage of errors was comparable to results obtained by children in a previous study. In other words, the evidence leans toward the cognitive load hypothesis. There was one difference between the performance of children and adult Lx learners: the latter benefited from referential cues (unlike children). Referential cues are found in the case of two referents (two frogs in our example): the fact that there is a frog on the napkin and one that is not clarifies which frog the sentence is referring to.

Captioning

The usefulness of captioning and subtitling,[13] that is, the presence of a transcription of the verbal component in a video on the screen, in Lx acquisition is "well established" (Gass et al., 2019, p. 86).

Various studies showed that learners could benefit from watching videos with subtitles, beginning with an early[14] study by Dow and Price, reported in Price (1983), which alerted researchers to the positive effects of using captioned video (see Specker, 2008; Montero Perez et al., 2013; Vanderplank, 2016; and Godfroid, 2020 for reviews of the literature). Montero Perez et al.'s (2013) meta-analysis on the use of captioning found that learners benefit from captioning in better listening comprehension (p. 728) and in vocabulary learning (p. 730), although it is unclear whether lower- or higher-proficiency learners benefit more (Montero Perez et al., 2013, pp. 722–3; see also Gass et al., 2019). They concluded that "captioning can be considered a powerful tool for reinforcing learners' comprehension of video materials" (p. 731) and "bimodal input fosters vocabulary learning" (p. 731).

The reasons that captions facilitate acquisition are numerous: among those mentioned in the literature are, first, obviously, they provide bimodal input, which is reinforcing (see Multimedia Learning section): seeing an orthographic representation of the words helps with segmentation and hence recognition of words; second, the relative permanence of the caption (which last up to 6 seconds) helps learners who struggle to keep up with the much faster and impermanent flow of speech; third, seeing the word may help students notice it and thus jump-start the acquisition process; fourth, captions may help disambiguate utterances; finally, the use of captions allows the use of authentic, interesting materials, such as sit-coms, as in Price (1983), which may improve motivation.

Eye-tracking studies of captioning and subtitling go back to studies such as d'Ydewalle et al. (1987) and d'Ydewalle and Gielen (1992), which showed that viewers automatically read the captions and subtitles even when they are in the same language, and thus are redundant, at least when they are presented at the normal speed (i.e., they last on screen and thus have to be readable in 6 seconds). Faster presentations degrade the performance of the speakers.

In a study by Muñoz (2017), the issue of whether the captions are in the same language as the video or in the L1 of the learners is examined along with the issue of proficiency. As we saw, proficiency is one of the variables concerned with learner characteristic, but the results are inconclusive. As Muñoz (2017) put it, "while studies on subtitle reading indicate benefits for L2 learning, there are still important gaps in relation to variables [such as] age and proficiency level" (p. 79). Her findings were that children (lower proficiency) skipped the subtitles less than adolescents and adults and spend significantly more time on them. This was true regardless of whether the subtitles were in English or not; however, beginners and intermediate learners spent more time on the captions when they were in English, showing that reading in Lx is more challenging.

Winke et al. (2013) examined the use of captions by speakers of several languages that use different writing systems. They showed that if the reading of the captions is too cognitively taxing, for learners of languages with writing systems different from the Roman alphabet (e.g., Arabic, Chinese), then the learners struggle to process the two modalities at the same time. More generally, they found differences in use of the captions depending on the language. Overall they find support for the hypothesis that captioning facilitates lexical learning and comprehension of the text.

Montero Perez et al. (2015) also tested the claim that captioning facilitates vocabulary learning using an eye-tracking setup. They investigated whether highlighting certain keywords in the captions or presenting just the keywords and whether the students had been told of an impending test had an impact on learning. The full versus keyword captioning yielded no significant differences. Knowledge of the upcoming test led to better results on meaning recall. Students in the keyword condition had longer fixation times but the knowledge of the test did not lead to any differences in initial fixation times. However, the students spent more time on the second pass (i.e., returning to the keywords for reanalysis). Montero Perez et al. (2015) conclude that prior knowledge of the test is enough to trigger a different task mindset (recall that Buswell and Yarbus had shown that people look at scenes differently depending on the task; see Chapter 1). The authors also find a positive correlation between overall fixation duration and form recognition of the words.

Gass et al. (2019) investigated captions as part of the **multimedia principle** (see Multimedia Learning section); learning may be facilitated by concurrent information. Processing written and spoken language at the same time (multimedia) is presumed to tax working memory and hence executive function (see Reference section). Another detail is that the captions may compete for attention with the other visual channel, that is, the video itself. Gass et al. measured working memory separately. Their findings show that captioning facilitates comprehension (measured by a comprehension test). Working memory is not very significant at a low proficiency but becomes fairly significant with more advanced learners. Generally speaking, captions proved more helpful for low-proficiency learners. Gass et al. (2019) conclude that

> captions help with attentional control. In other words, the physical presence of written material on the screen provides an attention-drawing device in a way that normal spoken text, for example, does not. This salient, attention-grabbing written information is useful particularly when other information, such as aural information, becomes too difficult (i.e., high cognitive-load bearing) and, thereby, not accessible. Evidence for this claim came from learners in our study with more limited [working memory] capacities who tended to use captions more. (p. 98)

Surveys of the use of eye tracking in SLA can be found in Abdel Latif (2019), Aryadoust and Ang (2019), Godfroid (2020), Godfroid et al. (2020), Roberts and Siyanova-Chanturia (2013), and Winke et al. (2013). Hayakawa and Marion (2020) is a recent review that considers eye tracking and other forms of imaging (e.g., fMRI, ERP).

Applications

Adaptive e-Learning

Adaptive e-learning refers to customized and possibly personalized computer-mediated educational tools. Essentially, it answers the age-old problem of textbooks, namely that they are by definition "one-size-fits-all." To some extent textbooks have always been customized, and thus we see beginner, intermediate, and advanced language books, or introductions to linguistics targeted at linguistics majors and nonlinguistics majors, or books for native and non-native speakers, and so on. However, categories such as "advanced" are necessarily reductive and cannot accommodate individual variation. Adaptive e-learning, because of the capacity of the computer to model the learner, offers the promise to deliver individualized personalized materials that evolve with the learner.

The number of publications about adaptive e-learning is quite significant. Several meta-analyses of e-learning found it to be effective. Fontaine et al. (2019) concluded that it is effective in teaching skills to medical professionals. Zheng (2016) found that self-regulated learning has significant positive effects.

Multimedia Learning

Multimedia learning refers to the use of different modalities in teaching. Usually, it concerns the use of text and images, but it can involve film, video, audio recordings, computer visuals, and more. Alemdag and Calgitay (2018) provided a useful survey of the research using eye tracking to assess the effectiveness of multimedia learning. Clark and Mayer (2016) is a hands-on manual that presents results as practical advice for electronic course/material designers. Among the findings is the "multimedia principle" (Butcher, 2014), which suggests using both verbal and visual modalities (text, preferably spoken, and image), over a written explanation alone. In particular, the multimedia principle works best for novices. Eye-tracking studies have confirmed that learners go back and forth from the text to the image (e.g., Holsanova et al., 2009) in order to integrate the information. Liu and Chuang (2011) found that subjects viewing educational material on a computer spent more time reading than looking at the images. They also found evidence of integrative saccades/fixations (e.g., when the text mentioned "clockwise" movement, the next fixation was on the image displaying such movement). Irrelevant images were found to attract attention and thus presumably to lead to distraction.

Another principle is the "**contiguity principle**," that is, placing related verbal explanations and visual elements near to each other. There is a significant literature showing that integrated presentation (text and image are merged) is more effective than separated presentation (text and image are both presented, but separately). Johnson and Mayer (2012), for example, recorded the eye movements of subjects looking at the same information (the mechanism of braking in a car) presented as text integrated in the image, text separated from the image (below it), text integrated and labeled, and image labeled but text in a legend under the image. The learners presented with the integrated text had significantly more saccades from images to text and vice versa ("integrative transitions"; p. 186) and presumably because of this outperformed the students who received the nonintegrated version of the materials. The presence of the labels on the image did not change the pattern just discussed. Conversely, the labeled image and legend presentation did not yield a significant difference in performance and showed a lot of fixations on the labels in the image, suggesting that the learners used the labels to access the legend information. However, it seems that introducing an extra step is a less efficient method given that under that condition learning was not as effective. Overall, the learners spent more time looking at the text than at the images.

The "**signaling principle**" (Richter et al., 2016) proposes that highlighting some information, or **cueing** it, supports or facilitates the integration process and hence learning. Signaling can be textual (e.g., deictics) or visual (e.g., arrows, contrastive coloring, etc.). Bolding and italicizing in text may be considered visual signaling. A text may signal important parts (e.g., key terms) or its organization (e.g., by providing titles of chapters and sections in larger and bolder font). In e-learning materials new approaches such as color cues that move in the image/text to highlight change have been used. Signaling is more effective in low prior knowledge, since presumably experts no longer need it. Indeed, Scheiter and Eitel (2015), in an eye-tracking study confirming the signaling effect, noted that false cues (i.e., items that deliberately misidentify correspondences between the text and the images) are attended to initially but successively discarded. Elements that were signaled had a significantly higher number of fixations. However, in the misidentified cues, the attention dropped over time and was switched over to the nonsignaled but relevant elements. Longer dwell time on cued features is a well-established result (e.g., Mayer, 2010, which reviews several studies).

Cueing is not only achieved by using arrows, typefaces, and/or color. At its most abstract, cueing is merely a manipulation of the frequency of some aspect of the stimulus. For example, if learners were to find that in nine cases out of ten important information was to be found in the top right-hand corner of the screen, they would adopt this strategy. The learners need not be aware of the statistical regularity they have found (implicit learning). As Sisk et al. (2019) put it, people find the correct button in a familiar elevator faster than in one they have never seen. This has been called "contextual cueing." There are complex interactions of location and other features. For example, Goujon et al. (2012) showed that a change in color scheme degrades the cueing effect. Other types of interactions are reviewed in Sisk et al. (2019). An interesting feature of cueing is that it appears to be "local" in the sense that nearby items trigger cueing but items that are far from the target produce less of an effect (Sisk et al., 2019, p. 3). Eye-tracking evidence shows a reduction of saccade number in repeated scenes/images, as well as a shorter search time, showing that the search is facilitated by the repetition.

Learning Strategies

There are a number of studies that study "learning styles" using eye tracking. For example, Tsianos et al. (2009), Al-Wabil et al. (2010), Mehigan et al. (2011), Koć-Januchta et al. (2017), and Luo et al. (2020) found a correlation between dwell time on visual content and visual learning style. However, the concept of learning style itself has been rejected by numerous scholars (e.g., Riener & Willingham, 2010; Rohrer & Pashler, 2012; Kirschner, 2017; Papadatou-Pastou et al., 2020) as essentially a "myth." The results previously reported may thus merely be describing a preference of the participants. For example, Mehigan et al. (2011) established that it is possible to detect the users' preferences in terms of verbal versus visual presentation and that number of fixations and dwell time on images correlate with a visual preference. This opens the potential for an adaptive presentation that delivers the information in the format favored by the user.

Conclusion

It is clear that the availability of the new technology of wearable, affordable eye trackers has made advances possible in the empirical study of classroom and educational interaction that were literally impossible two decades ago. The insights provided by eye tracking are beginning to impact teaching practices and also educational software and interfaces. In particular, real-time feedback systems capable of assessing the degree and direction of the attention of the learner seem to be pointing at exciting future developments.

Self-Study Questions

The following questions pertain to the contents of this chapter. You should be able to answer them without looking up the answer in the book.

1. Explain the RAGE hypothesis.
2. Is captioning useful for Lx learning?
3. How can mind wandering be remedied?
4. Explain the multimedia principle.
5. Discuss Godfroid's test of Schmidt's noticing hypothesis.
6. Discuss Winke's test of the "input enhancement" hypothesis.

Words to Know

intelligent tutoring	student model	mind wandering	mind probe
mindfulness	flow	adaptive e-learning	multimedia learning
multimedia principle	contiguity principle	signaling principle	cueing

Further Reading and Sources

For discussions of the applications of eye tracking to teaching, see Lai et al. (2013) and Jarodzka et al. (2020). For SLA and eye tracking, see Winke et al. (2013), and Godfroid et al. (2020).

CHAPTER 9

Applications in Related Fields

One may be surprised to find a chapter on applications of eye tracking *outside* the field of linguistics in a book about eye tracking *and* linguistics. However, in our experience, the "solitary genius" model of the scientist is slowly being replaced by the research team model. Increasingly, linguists are part of multidisciplinary teams of researchers who each contribute a different facet of expertise to a study, which each scholar individually could not realistically have the expertise to accomplish on their own. Teams routinely include statisticians (and many departments generously make available their expertise university-wide), psychologists, computer scientists, and of course linguists. In fact, in some subdisciplines of linguistics, such as forensics, the practicing linguist is necessarily part of a team of experts who work for a law firm. The same goes for linguists working in marketing and engineering environments.

This chapter then comes from the increasingly broad necessity of linguists to work with other scientists not only from neighboring fields, such as psychology or computer science, but also from fields that are much further out, such as engineering and marketing. Unless linguists know the kinds of research that are being performed on eye-tracking platforms and using eye-tracking methodologies outside of their direct field of expertise, they cannot know where else their particular kind of insight might be applicable. Language is ubiquitous. Computer programs have entirely linguistic interfaces, recently voice-activated, and even graphical interfaces often have linguistic aspects (e.g., pull-down menus). Marketing obviously relies on copywriters who produce text and are eager to know how text is processed by their (potential) clients and how text interweaved with images, buttons, and pull-down menus is processed on web pages. Even engineering environments, such as airplane cockpits, require labels and warnings that may be linguistic. Linguists have much to gain from being aware of the kinds of (potential) applications that exist in the research world outside of the boundaries of linguistics. Perhaps this chapter may earn you a job!

A less mercenary but just as compelling reason for being interested in the psychological problem of visual scene perception, for example, is that increasingly multimodality is being acknowledged as an important topic of research and obviously many cases of multimodal visual input will include text and a visual scene or text and a visual factor (e.g., pull-down menus in a computer interface). It stands to reason that the processing of multimodal items must include an understanding of visual processing. In what follows we will review some applications to various fields. It goes without saying that we make no claim of exhaustiveness. We selected these applications because we thought that they were most interesting or relevant for linguists.

Human-Computer Interaction

Eye tracking in human-computer interaction (**HCI**), that is, the interplay between computers and their users, has been used primarily either to assess and/or manipulate interfaces or to assess user attention, on-task status, and emotional involvement. We have examined the latter approaches in testing and education (see Chapter 8) and briefly discussed embodied conversational agents in our discussion of social eye tracking (see Chapter 7). We now turn to the uses of eye tracking in HCIs. An interface is usually what appears on screen, text, or, in the case of graphical user interfaces, icons, and so on. Of course one can also have audio interfaces (Alexa and Siri, among the more sophisticated, but beeps and other sounds are examples of those) and haptic interfaces (vibration, pressure, etc.). The earliest interfaces between computers and humans were punched cards or hardware switches. Technically, an interface is a symbolic layer between the machine (the computer processor, which takes orders in machine code) and the user; whether the user interacts with the computer via toggling switches, punch cards, typing text, or moving icons on a visual representation of a desktop, all those means of interacting are interfaces between the machine and human.

Eye tracking can be used in HCI essentially in two modes:

- **Diagnostic:** The eye-tracking device is used in a passive mode by only recording data that are interpreted at a later time.
- **Interactive:** The eye-tracking device actively provides input to another system (e.g., controls a mouse on a computer screen) in real time.

The diagnostic use is by far the most widespread, as most of what we have described in this book is diagnostic: data are collected and then later analyzed. However, most people associate the word "diagnostic" with the medical profession. Indeed, diagnostic uses of eye tracking are found also in the medical field, where eye-tracking data are used to diagnose schizophrenia, autism, ADHD, Parkinson's disease, and numerous other areas (for a survey of applications of eye tracking to the medical field, see Harezlak & Kasprowski, 2018). An area of particular interest in the medical field is the differences in patterns of gaze between experts and novices (Harezlak & Kasprowski, 2018, pp. 186–8).

Interfaces

In this section first we briefly review uses of eye tracking as part of an interface and then we focus on the use of eye tracking in augmentative and alternative communication (**AAC**) devices.

Gaze as a Pointer

There have been numerous studies on the use of gaze as a pointer. Essentially, the idea is to use gaze instead of the mouse. The user can then "click" by pressing a key on the keyboard or using another device. Another approach is to use dwell time as the trigger. When a certain threshold has been reached the computer interprets a sustained fixation as a click. An example of this application described by Jacob (1991) is a display of the location of various ships, such as a radar screen. When the user fixates on a ship, if they look then to a text window, they will find all the information about the ship they looked at. Some studies, reviewed in Jacob (1991), showed that activities such as cursor movement could be done faster using eye-tracking techniques.

There have been applications of eye tracking to controlling computers. A good overview can be found in Zammitto and Steiner (2014). Some recent video games include the option of using eye trackers. A good example is *Assassin's Creed Valhalla*, which can be played with a Tobii eye tracker, which allows the player to control the field of vision and to shoot at a point by looking at it. There are videos on YouTube that demonstrate these features and others.[1] Moreover, we might mention that some systems are being developed that allow entering a pin (e.g., for accessing a locked smartphone) using the user's gaze.

Computer Science

Sharafi et al. (2015) reviewed the uses of eye tracking in software engineering and found that it has been applied mostly to model comprehension, code comprehension, and debugging. Model comprehension concerns the way users use visual representations of the workings of the computer software (models). Experts are found to use information such as color or layout to process these representations faster than novices. In code comprehension, we find that code requires longer fixation times and more regressions than regular text, as one would expect. We also find the familiar pattern that experts and novices display different patterns of gaze. For example, novices spend more time reading the comments of the software, whereas experts focus on the meaningful parts of the code (Crosby & Stelovsky, 1990). Likewise in debugging (finding and correcting errors in computer code), high-performing subjects look more at the visual representations, rather than the actual code (Bednarik, 2012). Other studies have focused on how different styles of notation affect performance. For example, Sharif and Maletic (2010) showed that variables that use the "my_file" notation are processed faster than those written "myFile."

AAC Devices

In the case of AAC users, eye movements are used to control a computer and in particular to select/produce text. In some cases eye movements may be the only option available to the user. This has indeed been a central focus of the research since the early stages (e.g., Hutchinson et al., 1989).

When dealing with speakers who are unable to use their voice and who have limited mobility, controlling a speech-producing device with eye gaze is an effective solution and in some cases the only one. Essentially, the speaker selects by gazing at an area on the screen of the device and the software interprets the gaze in the same way as a click would be interpreted. Figure 9.1 shows how these devices are organized.

FIGURE 9.1 *A screenshot of an AAC communication device. SpeechWatch: Earable AAC device from Gus Communication Devices.*

A different approach is **eye typing** in which text is entered using the gaze as a pointer for typing (Majaranta & Räihä, 2002). This can be using a visual representation of a keyboard, which is activated by gaze (e.g., *Visioboard*; Charlier et al., 1997). Figure 9.2 shows what the Visioboard interface looks like. The interaction is fairly simple conceptually: the user focuses on one letter; the system gives some sort of feedback, for example, highlighting the letter; and the user confirms the choice and moves on to the next letter. The addition of autocomplete (i.e., predictive text) is obviously a boon in these situations. In some cases, users may have trouble keeping a steady gaze or focusing. In those cases a technique known as scanning is used: the screen is divided in a grid which is scanned in a given order, determined by the general direction of the eyes, so that only a vertical or horizontal direction needs to be identified by the eye tracker. Selection is done with a particular eye gaze, for example, off the board. Usually, the selection requires a threshold dwell time, to avoid the **Midas touch** problem (i.e., selecting everything that the user looks at; Jacob, 1991). In a different implementation of the same idea of threshold dwell time, in a translation program, fixating on a word longer than usual triggers its translation. We saw a similar idea in the tutoring software that defines difficult words for dyslexic children (see Chapter 8).

An interesting approach, developed by two researchers at Bauhaus University, in Weimar, Germany, is based on pie-chart-looking displays

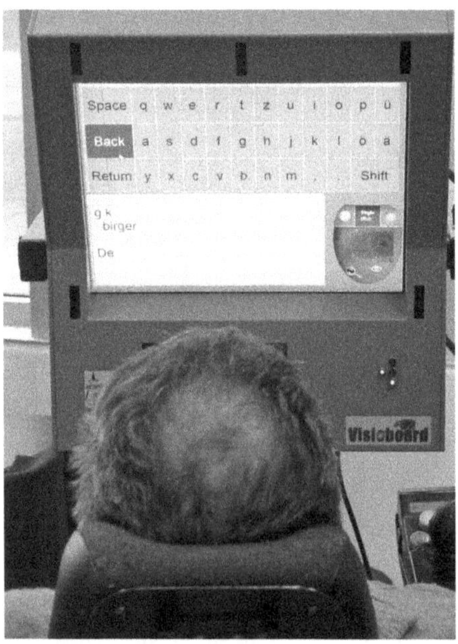

FIGURE 9.2 *A screenshot of a user interacting with the Visioboard. Reproduced with permission of COGAIN.*

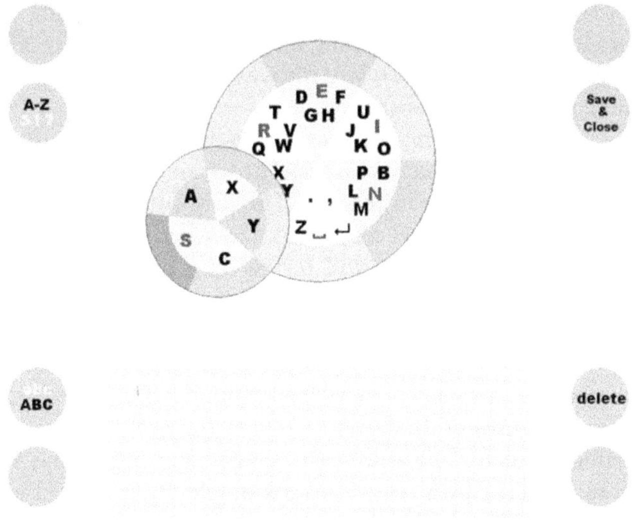

FIGURE 9.3 *The pEYEwrite interface (Huckauf & Urbina, 2008). Note the activated subpie "a, x, y, c, s." Permission granted by Anke Huckauf.*

(see Figure 9.3) and is named accordingly pEYEwrite (Huckauf & Urbina, 2008). When the user selects one of the six "slices" of the pie, another circular display "pops up" and a letter can be selected with one saccade. In order to facilitate usage, the most frequent letters are in red. A dwell time of 400 ms activates the selection. The text appears in the rectangular window at the bottom of the image. Other examples of eye typing are reviewed in Duchowski (2018, p. 62).

Typing and controlling a computer are not the only applications that have been implemented with eye control. EyeDraw (Hornof et al., 2003) is a program that allows the user to draw a picture controlling the drawing program with their gaze.

Visual Scene Processing

The study of the visual processing of images is not as advanced as that of reading. There are various reasons for this, but the most obvious one is clearly stated by Rayner and Castelhano (2008): while reading is a clearly goal-oriented task, that is, understanding the text, looking at an image is a much less defined activity. One can look at an image to remember it, to search for something in it, or just to enjoy its esthetic quality. Different goals will lead to different processing strategies, some of which may be idiosyncratic (i.e., developed by the individual subject for an ad hoc goal). For example, when searching for a small item in a crowded scene (e.g., in the *Where's Waldo?* series[2]) the viewer may use a grid-scanning technique starting from left to right and sweeping the image systematically left to right in a grid pattern. This is obviously *not* the way a viewer will scan a scene for an object under normal circumstances.

Generally speaking, reading and viewing an image deploy fairly different strategies. For example, during visual scene processing both the saccades and the fixations are longer than in reading. The perceptual span is also larger in image processing (Higgins et al., 2014). However, even more significant differences are found, which we will review very briefly in what follows. There is some evidence that early fixations are driven by visual features rather than by semantic features (Henderson & Hollingworth, 1999). As the exploration of the visual scene progresses, semantically significant areas are fixated more (so, e.g., a visually empty background will receive less fixations than, say, human figures in the scene).

The starting point for the study of visual processing is that the visual input is not random (i.e., white noise, think of television static) but consists of areas having different visual features. At its simplest level, imagine looking at a computer video image consisting of a left black side and a right white side. The contrast between a black pixel to the left of the boundary and a white one to the right of the boundary will be maximal, whereas there will be no contrast between two black pixels, for example. This is relevant because studies have shown that the gaze is attracted to areas of high contrast (e.g., Reinagel & Zador, 1999) as when pixels of the same color are grouped together (as in our extreme example, in which all the black pixels are on one side). From these two facts, it follows that viewers are attracted by borders between visual objects, since that will be where contrast and grouping correlate the most (Reinagel & Zador, 1999). This sort of analysis based on features of the visual input is a case of bottom-up processing. Other bottom-up salient factors are motion (which you will recall has dedicated neurons) and brightness. As the reader will recall from the studies of Buswell (1935) and Yarbus (1967) we reviewed in Chapter 1, top-down factors also contribute to the processing of images. For example, the task given to the viewers will affect what they look at. Top-down factors are volitional (i.e., controlled, at least in part, by the subject).

Saliency Maps

Bottom-up processes are integrated in the **saliency map** approach (alternatively spelled as "salience map"). The idea is that prior to reaching the threshold of attention, the visual system establishes a two-dimensional representation of the visual field, which encodes the saliency of objects in the map and thus determines the most salient one, which is then attended to by a foveal fixation. Itti and Koch (2000) implemented a saliency map, limited to such features as color, brightness, and orientation (up/down, sideways, 45°, etc.). The neural correlate of the saliency map is found in the superior colliculus (Chapter 3), an area of the midbrain, which contains a set of neural cells that control saccades and a set that stores candidates for the next saccade location.[3] An example of visual search is provided by Zelinsky et al. (1997) who studied the saccadic movements of subjects who had been presented an object (e.g., a plush toy) and were asked to find it again in another image in which it was found among other equally plausible objects (other toys) in a naturalistic background (a children's crib).

Note how in Figure 9.5 the first fixations tend to fall in the central area and by the third fixation arrive at their targets. Note that in Figure 9.4 one item was removed from the display (the fifth one, from left to right), as one

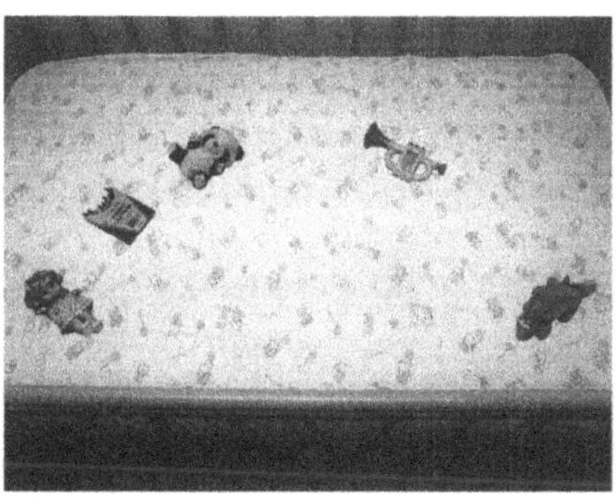

FIGURE 9.4 *One of the backgrounds in the Zelinsky et al. (1997) study. Reproduced with permission of* Psychological Science.

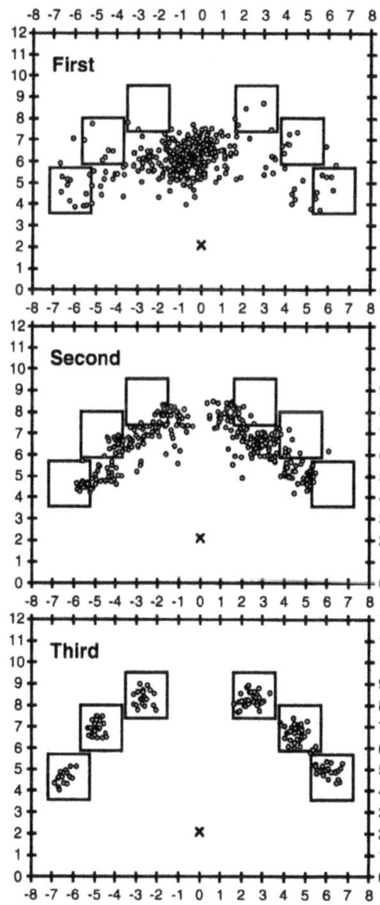

FIGURE 9.5 *The fanning out of the overall set of fixations in the Zelinsky et al. study. Reproduced with permission of* Psychological Science.

of the experimental conditions was the target search was absent from the scene. Figure 9.5 refers to all targets being present.

Far from sequentially scanning the objects, the participants used an approximation strategy starting more or less from the center of the visual field and moving out toward the targets, in a zooming fashion (see Chapter 4). This tendency of saccades to fall roughly in the center of a scene between possible targets has been called the "center-of-gravity" effect. It seems to be motivated by the fact that in the early phase of the search only very general, coarse information about the scene is available and the successive fixations refine this first approximation thus coming closer to the target (i.e., an object being searched for or a salient feature; Rao et al., 2002).

Top-Down and Bottom-Up Features

Top-down processes take longer (and hence will apply mostly after salience-driven bottom-up factors have applied) and involve shapes, semantic categories (e.g., "dog"), memory (has the subject seen the object before?), expectations (in an ice cream store if someone hands you something you expect an ice cream cone, not a can of soup), and finally background knowledge (schemata, frames, etc., i.e., encyclopedic knowledge, for example, of what ice cream cones look like). A good example of top-down influence on vision is that objects that are in an incongruous, unusual context grab the viewer's attention. So, a cow in a field with other cows would not be visually salient, but the same cow in a baseball field would be. Clearly, this sort of effect is determined by higher-level semantic expectations. So, in the earlier phases of processing, saliency determines the searches, as the outcome of bottom-up features. As the process continues more top-down processes kick in, such as expectations. The detection of changes in a visual scene (e.g., Spotorno et al., 2013) is a good example of the interplay of these processes. For example, it is easier to detect change in the consistent elements of a scene. Another example of top-down processing is the fact that in search tasks when the object is in an expected place (a person on a footpath, a car in a parking lot, etc.) it is found faster than when the object is in an unexpected location. Thus in a study by Neider and Zelinsky (2006) cars (which are usually found on the ground) were found significantly faster than helicopters (which can be either on the ground or in the air). Also, when the target was present, fixations were concentrated in the likely area (e.g., the sky, when searching for a blimp), whereas if the target was absent, the search was expanded to areas where one would not normally expect to find the item (p. 620).

The Gist of an Image

However, even with extremely short presentations (20 ms, i.e., less time than it takes to plan and perform a saccade, which then excludes that eye movements may be involved), experimental subjects were able to recognize whether the image displayed an animal or not (Thorpe et al., 1996). These kind of results have been seen as evidence that viewers capture the "**gist**" of an image[4] from the first fixation (Henderson and Hollingworth, 1999, "concept of a scene"; Oliva, 2005). Significantly, the gist of an image includes "all levels of processing, from low-level features (e.g., color, spatial frequencies) to intermediate image properties (e.g., surface, volumes) and high-level information (e.g., objects, activation of semantic knowledge" (Oliva, 2005, p. 251). Intuitively, the gist of an image would alert the viewer that the image is an indoor scene, or conversely, an outdoor landscape, but not provide enough detail to be able to tell what kind of room the scene is located in (kitchen vs. dining room, for example). An example of a model that integrates bottom-up and top-down features can be found in Torralba et al. (2006). One of the most peculiar findings about very short presentations of visual scenes is that while they allow for some semantic recognition and facilitate searches in a similar scene presented later, they do *not* allow the viewer to recognize changes (even very large ones) to the scene. If, for example, a white screen is inserted between a very short presentation and a modified version of the scene, the subjects will not detect the change. This is taken as evidence that further processing is required for the information to be stored in memory (Rayner & Castelhano, 2008, pp. 25–6).

Visual Processing Conclusion

So, in conclusion, besides the type of visual processing in reading in which the eye proceeds left to right (or in the direction of reading of the given writing system) with short saccades of roughly one word length, there also exist other modes of processing visual information. There is an extremely fast (less than 100 ms) apprehension of the gist of an image. There is a slightly slower process of bottom-up visual feature saliency processing, followed by an even slower top-down process. Finally, there is a different mode of processing called scanning in which saccades are broader and directed at headlines or pictures. Essentially, in scanning the viewer is trying to establish which areas will require deeper, detailed processing (Wartenberg & Holmqvist, 2005). In summary, the processing of visual scenes is a complex area, in which the research is still in flux, but we already know that it is quite different from the processes involved in reading.

There is some contrasting evidence about local and global exploration of visual scenes. There is evidence going back to Buswell's and Yarbus's groundbreaking work (see Chapter 1) that showed the task-related nature of visual exploration. Viewers start out by exploring the image globally (e.g., DeAngelus & Pelz, 2009; Tatler et al., 2010; see also Wade, 2015) and then narrowing down to a more detailed, local examination. However, there are also studies that show that users start from the center of the image and then switch repeatedly from local to global exploration, at least in the case of complex stimuli such as advertisements (unlike, say, a nature scene). For example, Liechty et al. (2003) state that subjects appear to start visual attention in the local state, with a quite long fixation, much more likely at the center of the stimulus. This, however, may not be strictly a "local fixation," in that it serves a different purpose. We conjecture that around the initial fixation, preattentive computation of perceptual salience occurs, based on features such as color, orientation, shape, and size, relative to their background (cf., Itti & Koch, 2000). Here, salient regions in the stimulus are identified (p. 533).

In other words, the participants in the experiment would be building their saliency map while fixating in the center[5] and would then start exploring other areas, based on the saliency map.

Face and Facial Expression Perception

The perception of faces has become a major topic of research in the study of vision. There are specific areas of the brain dedicated to the perception of faces. Recall that people suffering from prosopagnosia have a difficult time recognizing faces, and specific brain features have been associated with prosopagnosia (see Behrmann et al., 2011).

Yarbus's work on the prominence of the eyes, mouth, and nose, in facial scanning, has exerted a significant influence on the field. For example, Mehoudar et al. (2014) reviewed some of the studies which show a typical pattern of fixations on the eyes, nose, and mouth (see Figure 9.6) and the influence of task-based preferences (e.g., when asked to match facial expressions, participants focused on the mouth more than on the eyes, unsurprisingly). Mehoudar et al. (2014) also reported the presence of individual differences in the pattern of face scanning that is stable through time (the participants were observed twice at a distance of eighteen months).

It has been established that cultural differences also affect face scanning. For example, Blais et al. (2008) and Kelly et al. (2010) found that Westerners focused on the eyes and mouth and East Asians on the nose (central area). Haensel et al. (2020) reported that British/Irish speakers looked more at the mouth, whereas Japanese speakers looked more at the eyes and the nose. So we must conclude that, while facial perception is universally facilitated relative to the perception of other objects, culture modulates how the features required for **facial recognition**, and so on, are acquired, that is, by direct foveal fixation or in parafoveal (peripheral) vision.

A
B
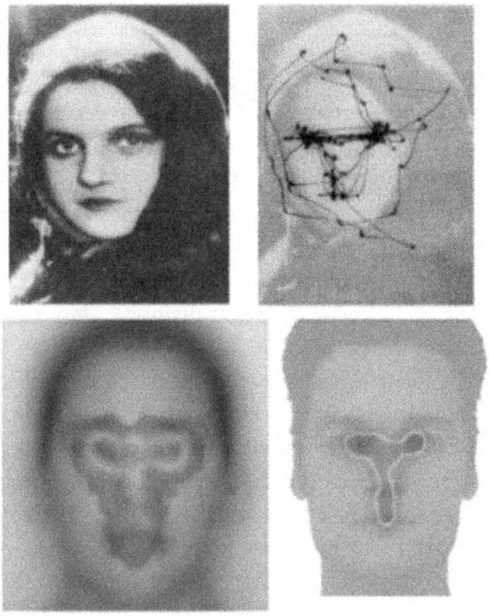

FIGURE 9.6 *The typical triangular pattern of fixations on the face. Yarbus' images are in the top row.* Source: *Mehoudar et al. (2014), reproduced with permission of* Journal of Vision.

An interesting phenomenon is the existence of the own-race bias, that is, a diminished capacity to recognize/recall the facial features of members of another race. This has been confirmed and its source (misallocation of cognitive resources) has been revealed by eye-tracking studies (Wu et al., 2012; Hills & Pake, 2013). Viewers display different gaze patterns when looking at faces of a different race. Wu et al. (2012) found "a more active scanning pattern during the encoding of own-race faces than other-race faces; with a higher number of saccades, and more frequent but shorter fixations" (p. 213). Generally speaking the processing of own-race faces is faster and less effortful, as confirmed also by pupillometry measurements, showing that "pupillary diameters were larger when viewing other-race than own-race faces, suggesting a greater cognitive effort when encoding other-race faces" (p. 202). Hills and Pake (2013) and Hills et al. (2013) showed that if the attention is redirected, using fixation crosses, to the physiognomic diagnostic features[6] of the face, the own-race bias disappears.

Finally, we discuss the perception of facial expressions and emotions. Affective computing, as this area of HCI has been referred to, consists of recognizing and/or reproducing emotional displays in interlocutors and, in the case of production, synthesizing them. The connection to the work on embodied conversational agents (see Chapter 7) should be obvious: a significant part of face-to-face interactions consists of synchrony, that is, matching the other speaker's gestures, facial expressions, and so on (e.g., Paxton & Dale, 2017). Behavioral synchrony has been shown to facilitate cooperation and social connection (Baimel et al., 2015). Thus since one wants the users to cooperate with the computer, recognizing and reproducing the user's facial expressions is essential.

There have been a number of promising studies that showed some correlation between eye-tracking metrics and emotions. For example, Raudonis et al. (2013) claimed 90 percent accuracy in identifying amusement using pupil size and motion. Rajakumari and Selvi (2016) used the measurement of the distance between sclera and iris to detect six emotions (anger, fear, happiness, sleep, focus, and disgust); they self-reported "good" accuracy (p. 76) but did not provide figures. Zheng et al. (2020) used pupil size alone to identify four basic emotions and achieved 57 percent accuracy. Lim et al. (2020) provided a good review of this fledgling field of research.

Lim et al. (2020) also reviewed several papers mostly dealing with pupillometry in combination with other physiological measures (such as EEG, electrooculography (see Chapter 1), etc.). However, these studies generally discriminated among very general categories of emotions (e.g., positive vs. negative) and achieved identification success rates in the high 70 percent range. A 2019 dissertation (Anwar, 2019; see also Anwar et al., 2018) achieved 93 percent accuracy and discriminated seven emotions. However, the eye-tracking component in this study is not related to the detection of emotion but is limited to tracking of the position of the pupil to determine gaze direction.

Engineering

Aviation

There is a very large number of eye-tracking studies of the gaze of pilots. Peißl et al. (2018) reported that an estimated 80 percent of flight information comes in visual form. This, and the high stakes of human error in flight, may explain the interest in eye-tracking technology in aviation, both civil and military. Indeed the earliest studies in the 1950s were started by the US Air Force.

Eye-tracking data are collected using head-mounted or wearable eye trackers, if the data are collected in actual flights, or using regular eye-tracking devices if the participants use simulators. The participants are subject to a variety of scenarios, such as landing, responding to emergencies, and so on. Often experts are compared to beginners. Mostly the studies focus on the order and length of fixation on various AOIs, since the general idea is that by monitoring various processes (speed, altitude, roll, pitch, etc.) the pilot is able to keep the plane under control. One of the AOIs is of course the external world (i.e., the pilot looks out of the window). Generally speaking pilots look more at the instrumentation, rather than the outside world. Expert pilots look more frequently at the instrumentation than novices, but their dwell time is shorter, because they can extract the required information faster (Ziv, 2016, p. 89). Interestingly, experts look more at an indicator when a correction is needed, but they also return to the indicators periodically to make sure the plane is on the right flight path. Under stress or high workload, dwell times increase, and particularly for novices (Peißl et al., 2018, pp. 103–4). An interesting use of the research has been in the development of better displays to present the information in the cockpit (Peißl et al., 2018, pp. 105–6) and the flying of drones. Systems that can detect fatigue or sleep deprivation have been developed; they use blink rate, pupillometry, duration of eye closure, saccadic velocity, and so on. Surveys of the field are found in Ziv (2016) and Peißl et al. (2018).

Driving

As we saw in Chapter 1, one of the first applications of head-mounted eye tracking was to driving (Land & Lee, 1994). With today's interest in self-driving cars, as one could have predicted, the interest in what drivers look at when driving has surged. Chapman and Underwood (1998) showed that when driving on a straight road, drivers focus on the **point of expansion**[7] and occasionally "look around" at road markers and the edges of the road. When the road is not straight, early research focused on **tangent points** (Land & Lee, 1994; see Chapter 1, for an example), but more recent work (e.g., Lappi et al., 2020) seems to favor the waypoint[8] hypothesis, as evidenced by compensatory eye movements, that is, eye movements that prevent the blurring of the field of vision due to speed (see Chapter 4). These movements have been recorded in walking, bicycling, and driving and are more compatible with the waypoint hypothesis. The obvious significance of this issue will not escape the reader: if tracking the gaze of a moving subject, compensatory eye movements will need to be accounted for.

A serious issue is the difference between experienced and novice drivers, as some studies, discussed in Chapman and Underwood (1998), show that novice drivers with less than a year of experience driving are 40 percent more likely to be involved in accidents. Experienced drivers look at areas where potential threats are likely to be located (e.g., intersecting roads, exits, etc.), whereas inexperienced drivers focus more on what is ahead of the car. Expert drivers use peripheral vision to monitor their lane keeping. Novice drivers have longer fixations, presumably due to the need to process the information. Novice drivers are not as good as experienced drivers in using their peripheral vision to keep track of road markers. For more references, see Rosner et al. (2019, p. 950.)

In Chapter 8, we highlighted the potential danger of mind wandering in driving. According to data reported by Strayer (2015) the number of all crashes involving inattention ranges from a *minimum* of 25 percent to a percentage as high as 78 percent. Beside mind wandering, another risk is the looked-but-failed-to-see phenomenon, which Rosner et al. (2019, p. 938) compared to inattentional blindness (see the "gorilla" experiment in Chapter 4). Essentially, the driver is actually looking in the direction of the car or pedestrian they hit, but they do not see them, because they are concentrating on something else. According to a study reported by Rosner et al. 10 percent of accidents could be due to looked-but-failed-to-see situations.

From the standpoint of safety, drivers should be continuously scanning the road ahead to identify potential obstacles and taking measures to avoid them. This state of alertness and attention is called **situation awareness**, that is, "knowing what is going on in a given situation" (Rosner et al., 2019, p. 936). Indeed, when situationally aware, drivers assess the road ahead and predict potential threats (e.g., a car approaching from an intersecting road).

However, these conditions do not always hold. Broadly speaking "drivers tend to steer in the direction of visual gaze and gaze in the direction they intend to steer" (Strayer, 2015, p. 425). When the cognitive load increases, they focus on the center of the road ahead, which may lead to ignoring potential issue in the peripheral vision (e.g., pedestrians, cyclists). This is called "tunnel vision." Expert drivers also use "**anticipatory glances**" toward potential hazards, such as a truck parked on the side of the road, which might hide pedestrians or another car coming from a side road. However, when they are distracted (e.g., by speaking on a cell phone), the number of anticipatory glances drops by 50 percent as well as the recognition and recall of objects (Strayer, 2015). Novice drivers do not display the pattern of anticipatory glances. The same pattern of deterioration of attention while driving under heavy cognitive load (such as talking on a cell phone) is also found in other situations: lane changes, left turns, and so on. Generally speaking, novice drivers fail to evaluate correctly the speed and distance of other vehicles.

Marketing

One of the areas that has been most open to and interested in eye tracking has historically been marketing and specifically advertising, because of the obvious opportunities for selling products to consumers when one knows what the would-be consumer is paying attention to. The use of eye tracking (and other physiological measures such as fMRI, ERP, etc.) has been called **neuromarketing** (Rothensee & Reiter, 2019). An interesting recent development in this kind of research is the possibility of using the users' own webcam on their computers or smartphones to track their eyes and calculate where on the screen they are looking (He et al., 2014). Once researchers get over the hurdle of convincing the participants to opt in,

Rothensee and Reiter (2019, p. 831) noted that this approach is much faster and significantly cheaper than doing lab-based research.

Eye tracking is ideally suited to study the multiscreen viewing, a recent trend which involves, for example, looking at one's phone *while* watching television. There are conflicting reports on the prevalence of this practice (Rothensee & Reiter, 2019, p. 843) with reports by users of multiscreen viewing ranging from 28 percent to 49 percent of the time. Rothensee and Reiter (2019, p. 844) explained this inconsistency by the supposition that users' reports are unreliable because users simply do not recall looking at their phone or tablet. The eye-tracking data show that users look at their digital media twice every minute.

Advertising

A common distinction in the study of advertising is between the visual elements (pictures), the written text, the **logo** of the company, the price, and promotional information such as special offers, discounts, and so on. In what follows we will adopt these distinctions, bearing in mind that from a strict linguistic standpoint, text is text, regardless of whether it is a price or a discount or the copy of the ad. Logos are obviously a mixture of textual and visual graphic components.

A consistent finding is that viewers will look more at bigger ads (Pieters et al., 2007). This is consistent with the finding that larger-sized objects in the visual field attract attention. Pieters et al. (2007) found a positive correlation between the size of ads in newspapers and attention: this is particularly true of images, but less so of the text, where they did not find a significant correlation (but see also Pieters and Wedel, 2004, where the relationship was inverted).[9] Brand, price, and promotional information also showed a correlation between size and attention. However, Pieters et al. (2007) also found that a significant factor is also how many other ads appear on the same page (clutter): ads that are more distinctive get more attention. Pieters and Wedel (2012) show that ads can be recognized as such and the product recognized with a single fixation as short as 100 ms. Essentially, the viewers extract the gist of the ad in a very short time, a claim consistent with the research on gist (see Visual Scene Processing section).

However, not all images are the same: for example, Nielsen and Pernice (2010, p. 213) reported that viewers do not look at 85 percent of stock images and that simpler, less crowded images with a single subject are looked at more frequently. Viewers also show a preference for images that are related to the content (pp. 218–19), especially if they are clear, not blurry, and have high contrast. For images of people, the preference of viewers goes to attractive, smiling subjects, who are facing the camera, are authentic (i.e., not models but normal people with imperfections), and as we saw with an uncluttered background (p. 230). Also, images of one person are preferred over images of several people. Of course, images of dogs draw the users in (p. 241).[10]

Readers also look a lot at brands: in fact brands have the highest fixation-to-space ratio of all the elements, suggesting that while a larger image may attract attention a smaller brand does just as well. Brand images are also more memorable. Graphics, color, text, complexity, and the degree of clutter all influenced attention to advertisements (Ladeira et al., 2019). Finally, we should mention that in advertising research, the preferences and attitudes of the participants and most importantly their intention to purchase the item advertised are important variables (Ladeira et al., 2019).

Indeed, an important aspect of advertising, or one might argue the most important aspect of advertising, is choosing between competing options in the decision-making process that leads to a purchase. The **gaze cascade effect** is relevant to this process. Shimojo et al. (2003) described a cascade effect of gaze length: when two items are being compared, gaze length progressively increases on the item that is eventually chosen. Atalay et al. (2012) showed that a gaze cascade effect takes place in favor of the central item of a 3 × 3 matrix of products, leading to the choosing of the central item more frequently.

Wedel and Pieters (2008) elaborated a theory of visual attention in marketing. It incorporates the well-known distinction between top-down and bottom-up factors: consumer goals (e.g., saving) are an example of top-down factor, whereas the visual features of the ads (size, color, etc.) would be bottom-up. Their model incorporates the idea of the saliency map (Saliency Maps section) and of the gist of the image (The Gist of an Image section). So, attention is a function of top-down and bottom-up features: the top-down features (goals, expectations, memories of what a product looks like, etc.) determine the degree of informativeness of the image (if one is looking for Coca-Cola, for example, the logo and red packaging will be maximally informative), whereas the bottom-up features, integrated in the saliency map, will make sure we notice the red package.

Point of Purchase

By "**point of purchase**" we mean advertising that is located in the store where the would-be buyer is physically located. A rather surprising fact is that a large percentage of purchases are unplanned ("**impulse purchases**"); various studies report that 50 percent of purchases are made in the store without prior intention (Abratt & Goodey, 1990). Huddleston et al. (2015) reported percentages of impulse purchases as high as 68 percent.

An early result is that items on display sell more and that items at eye level are bought more frequently. In-store marketing includes signage, the labels and brands on the products themselves, and merchandise displays (i.e., the shelving in which the products are displayed). Obviously, the use of fliers, mailers, coupons, and printed ads in newspapers could also be included in point-of-purchase marketing, but they fall under the advertising heading. The importance of displays in marketing cannot be overstated:

> Merchandise displays are ubiquitous in retail settings. Retailers rely on displays to be silent salespeople, to draw consumers into the store and motivate them to touch, evaluate and purchase products. Displays have the capacity to increase sales, for example, [Nordfält] (2011) found that disorganized displays, which signal cheaper merchandise, can increase sales by over 900 per cent. (Huddleston et al., 2015, p. 562)

Nordfält (2011), using visual observation to check where customers look, also showed that digital displays increase sales as did displays in which the product is organized by color in vertical stripes (see Figure 9.7 for an example).

However, Huddleston et al. (2015) noted that more is known, in marketing, about the top-down factors (such as the customers' characteristics and

FIGURE 9.7 *Fruit arranged in vertical lines differentiated by color in a grocery store.* Source: *Carlo Martin Alcordo.*

goals: e.g., customers fixate more on what is important to them) rather than the bottom-up factors (properties of the stimuli).

The potential for eye tracking in marketing is remarkably broad. There is definitely a correlation between visual attention (fixations, dwell time) and brand choice (Pieters & Warlop, 1999). Remarkably, Rothensee and Reiter (2019) reported that in one study 64 percent of German consumers could not distinguish one brand from another (do you know the brand of dishwashing powder you use, or do you identify it by color and shape at the store?).

Chandon et al. (2009) used eye-tracking equipment to record the gaze patterns of participants looking at planograms (i.e., an image of a typical shelf, packed with goods (soap and pain killers, in this experiment) in various configurations. They stated that they reproduced the images in life size. One of their first findings was that self-reports (i.e., asking the participants to report where they looked or what they were interested in) are not reliable (p. 2). This highlights the importance of eye-tracking studies for marketing.

Usability Studies

Usability studies are focused on pragmatic goals: for example, in the case of web pages, does the web page succeed in doing that for which it was created? Usability applies also to interfaces (e.g., graphical interfaces vs. text-based) or the details of it (e.g., pull-down menus). Obviously usability goes beyond HCI, and so it broadly encompasses design, ergonomics, and more generally the affordances of an entity. Generally speaking, by usability we mean that something is efficient (i.e., easy to handle/maneuver to achieve the goals of the user), easy to learn, and pleasant to use.[11] Usability also includes **accessibility**, that is, the capacity for a person with a disability to get the same or a similar experience as other users. Thus, the usefulness of eye tracking in usability studies is fairly easy to establish, on the basis of the eye-mind theory: since the eye tracker can tell us what the user is looking at, and we can assume that that is what the user is thinking about. In this review we will focus on web usability and software usability. We will not focus on usability in the real world, so to speak. However, we considered point-of-purchase sales in the point-of-purchase section.

Generally speaking, eye tracking affords a greater degree of precision and analyticity in the study of users' behavior that other protocols (e.g., the think-aloud protocol, in which participants are asked to verbalize their thoughts while they perform a task). Specifically, they afford the capacity to tap behaviors below the threshold of consciousness. The kind of questions that eye tracking can answer uniquely (or just better than other techniques) are, for example, how many times did the user look at a button or link before they clicked on it? Did users have to reread or recheck the content? How long did it take for the user to notice (fixate on) a given item? Also, speaking generally, eye-tracking studies are often combined with more traditional studies (e.g., surveys). See Olmsted-Hawala et al. (2014) for more examples.

Heat maps and gaze paths are the two most frequently used techniques in usability, presumably to accommodate nonacademic clients who are paying for the study.[12] Although Rothensee and Reiter (2019, p. 827) reported a shifting trend toward more numerical presentation of results: "with repeated exposure to heatmaps, clients' interest shifted focus increasingly to numerical and comparable results. This also demonstrates that the purchasers of eye tracking research are now much more informed on how to interpret its results than at the end of the 1990s." Nielsen and Pernice (2010) noted that whereas software usability is primarily interested in optimizing the user experience, web usability is primarily focused on the practical goal of the website and so focuses on the goals of the users, be it buying something or finding information. So the pragmatic, practical aspect of usability studies should always be kept in mind, although the best usability studies easily meet, or exceed, the standards of academic studies. For example, Nielsen and Pernice (2010) report a study with 300 participants, a number larger than most academic studies.

One of the first findings that Nielsen and Pernice (2010) reported is that users look at web pages in a variety of ways, dependent in part on the task. Other factors affecting the order of processing are the layout of the page and the content. There is no fixed order of processing like in reading. This is similar to the processing of visual scenes (see Visual Scene Processing section). However, users are aware of some conventional positions for elements of web pages and their scan paths are affected by these expectations. For example, users expect web pages to have a menu (a list of options available on the website) that is distinct from the background page. Logos tend to be in the upper-left corner, whereas search buttons tend to be located on the top right of the page (Nielsen & Pernice, 2010, pp. 64–5).

Nakamichi et al. (2006) considered web pages, when looking for information (unfortunately, they do not report what web pages or information were used). They found a correlation between usability and length and speed of the saccades (in other words, when the page is hard to use, as per the subjective perception of the users) they find faster and longer saccades. When accessing a web page for a specific function (e.g., searching for information on the web), users focus only on the search field and ignore the rest of the page (Nielsen and Pernice, 2010, p. 338).

The phenomenon of **banner blindness** puts in particular relief that users are not merely reacting passively to the properties of the stimulus. In fact, web page users deliberately avoid looking at banners because they know they contain ads and are distracting from the goal for which they are scanning the page. Stenfors et al. (2003) reported that users having become aware of the usual location of ad banners in websites learn to avoid looking at them. However, the literature is mixed on the subject. Barreto (2013) and Pasqualotti and Baccino (2014) reviewed several studies with contradictory findings on banner blindness. For example, Hervet et al. (2010) reported that 82 percent of their participants fixated on at least one out of four banners. It appears that banner blindness is strongly influenced by the position of the banner and by the degree of expectation (i.e., the phenomenon disappears if the distracting stimulus varies randomly; Theeuwes and Burger, 1998). What is interesting here is that users actively avoid content that they can predict and dislike and while they may be unable to "weed out" unwanted unexpected stimuli upon first encountering them, they soon build an attentional "immunity" against the distractor. Drèze and Hussherr (2003), who also documented banner blindness (a banner has a 50 percent probability of being fixated; contrast this figure with 97 percent of TV ads!), suggested that processing of the ads may occur at a "preattentive" level, since banners are shown to affect brand awareness and recall.

The relatively recent mainstream adoption of online education has created a new field called **learning analytics** (e.g., Nistor & Hernández-García, 2018), which is concerned with the study of the interaction between learners and online educational platforms, both in the senses of Massive Online Open Courses (MOOCs), such as those offered by Coursera, Udacity, or edX, and of the use of online tools to deliver (more or less traditional) online courses.

We will focus on a topic that has recently become very pressing, that is, the use of learning management systems (LMSs) for instruction in college and graduate-level courses. An LMS is a software platform designed to offer an integrated online learning experience for students/faculty. It typically hosts course materials (textbooks, slideshows, videos, audio files, etc.), a grade book, a set of assignments or tests to assess the learning of the students, and discussion/interaction spaces, where students and faculty can participate in various interactions, including email. Other tools available may include wikis, rubrics, automated grading, automatic attendance/activity tracking, and so on. Generally, LMSs are fairly customizable by the faculty.

Conley et al. (2020) examined the usability of LMS by students. They noted that usability correlates with a positive experience, retention, and performance. Specifically, they contrasted a chronological layout of the course (week-by-week organization) to a functional layout in which materials were organized in assignments, discussions, and syllabus. Both layouts shared grades, tools, and announcements. The chronological layout did not have a syllabus slot visible in the home page. Conley et al. tracked time to first fixation and found unsurprisingly that the only significant difference was that the participants in the functional group found the syllabus faster than those in the chronological group. Participants found the chronological organization easier to use, but the difference was not significant. Likewise, the chronological organization was more efficient for submitting assignments and posting discussions. All the participants (n =

28) had prior experience with the LMS, so the effect of a novel environment was well controlled. While the differences between the two layouts were not significant overall, some helpful guidelines emerge, that is, that instructors should locate important features such as the syllabus at the top of a menu (p. 381) and that mixing chronological and functional approaches is probably a best practice of LMS course design.

Conclusion

It should be evident that eye tracking can provide substantive contributions to many areas of potential interest to linguists. This is certainly true, for example, of the consistent differences found in the gaze patterns of novices and experts. The reader will recall that we have encountered this difference in as disparate fields as driving a car; piloting a plane; the medical field, for example, in the examination of radiographies; and computer programmers. Indeed, the phenomenon is widespread (see Gegenfurtner et al., 2011, for a review). This line of thought may perhaps be integrated fruitfully with the study of Lx acquisition.

As we argued in the opening of the chapter, these applications are all potentially tied to linguistics because of the ubiquity of language and its role in mediating communication, even in nonlinguistic systems. To be sure, some applications are more obviously close to linguistics and some are further away, but we hope to have shown that the applications of eye tracking are many and quite widespread. It will be up to the ingenuity of the single linguists to find all the connections that are potentially there.

Self-Study Questions

The following questions pertain to the contents of this chapter. You should be able to answer them without looking up the answer in the book.

1. What is the difference between top-down and bottom-up factors in image (visual scene) processing?
2. What differences in patterns of gaze appear across activities in experts and novices?
3. Describe the solutions to the Midas touch problem in eye-tracking interfaces for AAC devices.
4. How would you explain the finding by Nordfält that products organized in vertical stripes of different colors attract the attention of customers? (hint: reread Visual Scene Processing section)

Words to Know

HCI	AAC	eye typing	banner blindness
interactive	diagnostic	usability	learning analytics
accessibility	gist	saliency map	vanishing point
anticipatory glances	point of expansion	tangent point	facial recognition
impulse purchase	point of purchase	logo	situation awareness
gaze cascade	neuromarketing	Midas touch	

Further Reading and Sources

There is no outstanding comprehensive overview of visual processing. On facial perception, a broad overview is Calder et al. (2011), which does include some eye-tracking work. On HCI, Jacob and Karn (2003) is a very good synthesis. On driving, see Rosner et al. (2019). On usability for websites, both Nielsen and Pernice (2010) and Olmsted-Hawala et al. (2014) are very good: the latter is more introductory. A survey of eye tracking in advertising can be found in Higgins et al. (2014) and Wedel (2015).

Part III

Using Eye Tracking

CHAPTER 10

Planning an Eye-Tracking Study

Writing a paper is the same activity, whether it includes an eye-tracking component or not. So, many of the things we will review here will seem familiar, from previous writing experiences. In what follows, we will consider some of the advanced research and writing skills needed to produce a publication-grade paper using eye-tracking **methodology**. What we will focus on here are the issues that are unique to, or at least more likely to be present, in an eye-tracking study.

Let us start by noting that there are many different ways, all equally valid, of setting up a research study. The main distinction is between quantitative approaches and qualitative[1] ones. Eye-tracking studies tend to be quantitative and so we will concentrate primarily on quantitative methods and particularly on the experimental approach, in which the experimenter controls (i.e., systematically changes) one variable and observes how the other variables react to that manipulation. When the experimenter cannot manipulate the independent variable, for whatever reason, we are dealing with an observational study. For example, Christensen et al. (2014), aside from the experimental approach which is typical of the sciences, listed among observational studies:

- The correlational approach, in which a relationship is established between two observed variables: for example, the age of a child and the number of words they know;
- The natural manipulation approach, in which the experimenter cannot manipulate the different conditions and so just observes them: asking participants to perform a task on the internet would be an example;

- Cross-sectional studies, in which participants are observed at a point in time, to determine their condition (thus, the mind probes used in the study of mind wandering are cross-sectional studies); and
- Longitudinal studies, in which participants are observed repeatedly over extended periods of time (months or years).

For those readers who come from the humanities, we will include some discussion of how to set up and execute an empirical study, but obviously we cannot go into sophisticated test designs. An introductory textbook on research and experimental design, like Christensen et al. (2014), is over 500 pages long! All we can hope to do here is to give the reader some pointers in the right direction.

Naturally, the idea of doing research starts from a question: for example, we may wonder can I show that lexical activation occurs before a given time frame? Or how much do readers look at the captions of images in a text? Or do people avert their gaze when they are lying? Or any other question along those lines will do. The exact question does not matter much, because the idea that you start from will almost certainly change and by the time you reach the final stages of planning your project it will have evolved. That is one of the ways that reading the relevant literature is important: it helps you refine your idea.

The Literature Review

The purpose of the **literature review** is to list all the relevant studies on the topic about which you are writing, review and synthesize their findings, and where necessary identify any gaps in the literature. In a sense, it is doing your "homework." The way to produce a literature review has changed significantly in the past ten to fifteen years. Generally speaking, it is now much easier to do so; however, there are a few caveats that are worth presenting. In what follows we will assume for the sake of the argument that you don't know much at all about the subject you want to research. This is obviously not likely, in the sense that you have probably taken a course or two on topics that include or are related to your topic. However, by assuming no prior knowledge we will be able to outline the research process more clearly.

The best place to start is an academic search engine, such as Google Scholar (`https://scholar.google.com`); Semantic Scholar (`https://www.semanticscholar.org`); Refseek (`https://www.refseek.com/`); Core (`https://core.ac.uk`), which specializes in open-access articles; the German-based Bielefeld Academic Search Engine (BASE: `https://www.base-search.net`); and Microsoft Academic (`https://academic.microsoft.com/home`), which offers a broader set of reports and visualizations, which include attempts to assess what subtopics are popular within a field, what journals and institutions are central to it, and so on.[2] These resources have a very broad coverage and are free. You input a couple of keywords, such as "eye-tracking automobile" and you get thousands of results.

There are other catalogs of scholarly sources, arguably better than free resources such as Google Scholar, which is error-prone,[3] since the references are computer-generated, but they are proprietary (i.e., they will charge you to consult them). However, a good research library should have access to at least some of them. The *Web of Science* is particularly good. *WorldCat* will locate any book, worldwide. You should familiarize yourself with your library's resources.

Academic search engines have another feature that makes them indispensable: at the click of a button they can tell you which other sources have quoted that particular source. This makes assessing the impact of papers, finding discussion and refutations, in short, building your literature review, much easier. You should take full advantage of these features. However, we come to your first caveat: electronic databases are based on and calculate the impact factor of an article or book (i.e., how many times it is cited). They tend to present the articles in the order of ranking (i.e., highest-ranked articles come first), which means that they may completely miss less frequently quoted, but equally valuable articles. So you should go beyond the first couple of pages of hits. To put it differently, you need to dig deep.

Literature reviews are done using a bibliographic format. The format is a set of conventions that help scholars know where to find the information. Consistent placement facilitates retrieval. In this book we use the American Psychological Association (APA) but you need to be aware of the fact that some journals require a different style or even have an in-house format. For example, Oxford University Press insists, unsurprisingly perhaps, on the presence of the "Oxford comma" before the last element of a list.

The work of scholars is further facilitated by software that allows you to store the articles you found online and automatically generate bibliographies (e.g., Zotero). The caveat here as well can be best described "garbage in garbage out": if the reference is not correct, and trust us, many are not correct, your bibliographic entry will be faulty. You should always manually check the references. Our next caveat is much harder to combat: suppose you have twenty or thirty articles on studies using the visual world paradigm. It is tempting to list them all in a nicely alphabetized list, as the APA recommends. After all, isn't it a sign of good scholarship that you read all of these articles? Also, since you are quoting all of these people, there is less of a chance that one of the referees of your paper will reject it because you did not quote them. However, the so-called "Google effect" (i.e., where students gather large amounts of information that they do not process and hence do not learn from; Yacci & Rozanski, 2012) is a real danger. If all you do is gather sources and report their existence, you have *not* produced a literature review. If you don't understand that the Tanenhaus et al. (1995) article is much more important (and should be quoted first) than any article we (alphabetically advantaged, Attardo and Pickering) could whip up using their methodology (and that APA would have you quote before Tanenhaus et al.) then you have not really mastered the field. A good literature review does not merely report the existence of a field, it engages with it. Please note that our point is not merely that you should list important contributions first, but that you should *discuss* them and place them in their historical and scientific context.

What Are You Looking For?

The obvious purpose of the literature review is to find out what scholars have said about your topic. This is the standard view of a literature review and there is nothing wrong with it. However, there is a second less obvious purpose in doing a literature review and that is to see what methodological tools are commonly adopted by scholars in your area, including the kind of statistical tests to be adopted. Another important factor, which you will learn about in the chapter on statistics (Chapter 11), is the effect size to be expected from a study. This will be crucial in planning your statical analysis. In short, one does not do a literature review just to be able to report what findings have been established in the field, but one does it *also* to find out how people in the field do research: what methodologies they use, what is the typical number of participants they use, the effect size they find, the kind of questions they ask. Being aware of all these aspects of the research will make you a better researcher.[4]

A recurring question from novice scholars is "how many sources" should they read. The answer, frustratingly, is that it depends. In some areas there are not really more than a handful of studies, whereas in other areas there are so many studies that reading them all would be literally impossible. As one student of ours once plaintively said: "everybody has written about my topic!" If you find yourself in a similar situation, a good heuristic is the law of diminishing returns:

- when each article your read increases your overall knowledge of the field only marginally,
- when all the articles quote the same sources and you have read all of them, and
- when most of the articles you find are replications of a set of original studies,

then it may be time to stop your search for sources. Needless to say, your supervisor, if you are writing a dissertation, is the one who will tell you when you are ready. Many researchers write up the literature review of their study first and so you may be able to give that part to your supervisor or to a trusted friend to assess.

Finding a Topic/Gap in the Literature

The literature review has another goal: finding gaps in the research. The reality of academe is that a study that replicates a previous finding is far less likely to be published than a study that makes a new claim on a topic that has not been covered before. While some have argued that this is problematic for the advancement of science and that more replication studies need to be produced, it remains true that novel findings are easier to publish. A novel finding is by definition something that has not been done before, that is, a **gap in the literature.**

A common metaphor is that of a funnel approach: you start from a topic within a field, for example, turn taking. Obviously, a whole lot has been written about turn taking (see Chapter 7 for a short overview). So you need to narrow it down, going down the funnel, to the role of gaze in turn taking. This is a reasonable topic. At this point you will start your literature review. Once you have read and annotated all the sources you gathered, you will become aware that some areas have been researched fairly thoroughly (e.g., gaze aversion as a sign that the speaker wants to hold the turn) whereas other areas, for example, gaze in multiparty conversations, have received much less attention. Indeed, Auer (2018, 2021) claimed that his work on three-party conversations is part of "a very small number of studies" using eye tracking in naturalistic environments in multiparty conversations. At this point, if you further narrow down to a special population, for example, speakers, you are almost guaranteed to be producing the first ever study on the subject.[5]

Research Questions

On the basis of your literature review and of the gaps in the literature you have identified a topic, for example, to carry on the example in the prior section, gaze in multiparty conversation among English as a Lingua Franca (ELF) speakers. Now you can start working out your **research questions**. The research questions will shape your methodology and are the core of your study, so it's important to get them right. Your literature review should have familiarized you with the kind of research questions that scholars in your field use. Often a small but significant tweak of a prior research question is all that is needed: for example, as we said applying the multiparty gaze in conversation paradigm to ELF speakers is relatively straightforward: all the questions that were relevant in the paradigm when applied to native speakers remain valid: for example, do speakers avert their gaze when they do not want to relinquish the floor? Are there signals that the speaker intends to relinquish the floor? However, other questions now become just as relevant: some are obvious, such as, are there differences between the gaze behavior of NS and NNS speakers? Others are subtler: are there differences in the behavior of NS and NNS speakers when interacting in a NS–NS and NNS–NNS context as opposed to a NS–NNS context? As you can see, none of these questions were relevant or even possible when all we had were NS conversations. A small tweak generated a host of original and interesting research questions.

Generally, the research questions are then formulated as hypotheses. A hypothesis should be as clear as possible and should state clearly what would count as a falsification of the hypothesis. So, for example, "there are differences between NS and NNS gaze behavior" is not a good hypothesis, whereas "NS will avert their gaze significantly more often than NNS when they want to hold the floor" is much better because it clearly implies that, if the NNS avert their gaze under the same circumstances equally or more so than the NS, our hypothesis is wrong, that is, has been falsified.[6]

Let us stress that, if you are a PhD or MA student, you should clear every step of the way with your supervisor. The job of supervisors is to help you craft a feasible project. The problem is that you need to walk a fine line between an overambitious project, which you will most likely be unable to complete successfully, and a project that is so specific that it becomes uninteresting (e.g., because your data cannot be generalized at all outside of the participants you used). Your supervisor will also know the practicalities of collecting the kind of data you need. For example, when planning an eye-tracking study, the time it takes to collect a 10-minute usable sample of gaze is much more than 10 minutes.

Only experience can tell you what works and what doesn't. A literature review is also in part a distillation of the collective experience of the field. What works get published? The experiments that failed, the projects that never made it do not get published, usually, and so you will never read about them. The best way to ensure that your project succeeds is to have a good methodology.

Methodology

So, you have your research questions; now the problem is, how do you find out the answers to these questions? The first step is to operationalize your definitions. What do we mean by "gaze behavior"? What AOIs will you be using? How will you define "relinquishing the floor"? How will you define "gaze aversion"? Obviously, the literature review will have some answers, but for some of the concepts you will need to come up with a definition. Once you have defined gaze aversion, for example, as a fixation in any AOI except the face of the other speaker(s), then you can decide whether you will be conducting an experimental study or a naturalistic observation. How will you collect your data? How will you recruit the participants? All of these issues need to be considered ahead of time and again, most of the time, the literature will have examples of other scholars who have solved that problem or a very similar one. The most important question, which should always be on the back of your mind in any decision you make, is "will this experiment, in this configuration, with this methodology, answer the question I want to answer?" You should never lose track of that fundamental goal.

A major part of the methodology is to decide which metrics you will use. It is tempting to think that you will just collect data and then see which metrics look good. Exploratory studies may be OK, but generally speaking, it is preferable to be selective and to only use those metrics that we expect to be relevant to the question/hypothesis at hand. Orquin and Holmqvist (2018, p. 1647) discuss the pitfalls of "fishing for significance." We cannot go into the details of the various metrics, but it is imperative that you thoroughly research all the metrics you plan to use. Sources such as Holmqvist and Andersson (2017) include detailed discussions of most metrics. See also Orquin and Holmqvist (2018) and Carter and Luke (2020), who reviewed some of the potential problems. Furthermore, the literature review will also alert you to potential problems. As we said, even a representative discussion of methodologies would be too broad for a book like this one, but we can use an example from Orquin and Holmqvist (2018) who showed that "total dwell time" is a problematic metric (Orquin & Holmqvist, 2018, pp. 1650–1) and should be avoided.

Finally, the methodology that you choose will likely have an impact on the statistical tools you will need in order to be able to analyze your data. We turn to the statistical analysis in the next chapter, but one of the least intuitive but most useful things to keep in mind is that you should know what kind of statistical analysis you will be performing *before* you finalize your study or submit it to your ethics committee board. So, in a very real sense, you should read the present chapter and the next one in one sitting, or at least not firm up anything until you've read the next chapter as well.

As we said, it may be counterintuitive, but the choice of statistical analysis may determine the kind of questions you ask or the way you categorize your data. For example, if you have quantitative data (such as raw test scores, for example) you can perform a Pearson correlation, but if your data are pass/fail you will need to perform a logistic regression or a polychoric correlation. You should not worry about what any of those statistical methods are (right now, at least), as that's not the point here, but rather you need to realize that whether you use raw scores, letter grades, or pass/fail will affect the type of statistics you can/should use. You should be aware of these issues *before* you collect your data, as once you have collected your data if you find out that the best test for that kind of data and/or question is one that requires advanced statistics you may have to do some extra studying.

The Planning Phase

Every good study begins with serious planning. The tricky part of planning an experimental study is that you want to manipulate the variable (e.g., the frequency of the words presented to the readers in your experiment) and not other confounding variables. For example, a well-known phenomenon is that participants get tired after a while or conversely that they learn from the previous questions what the task they are supposed to perform is and they do it better. Obviously, that's a problem because we are now manipulating not only the frequency of the words but *also* their position in the stimuli we present. There is a solution, which is to randomize the order in which we present the stimuli so that now if there is a fatigue effect, it will affect the words randomly and not always the same ones. This is just an example. There are literally dozens of such problems and solutions that are related to different experimental designs: if you are interested in two variables, this is called a factorial design, because you have two factors affecting the variable. Generally, this is visualized as a 2 × 2 grid. Let us use, for example, the Conklin et al. (2020) study we saw in Chapter 8 in which they compared L1 and L2 readers reading a story with and without the audio version playing. This gives us a 2 × 2 factorial design (see Table 10.1)

Another decision that has significant consequences is whether your participants will see all the stimuli or whether one group will see a fraction (e.g., half) and the other group the rest. If all your participants see all the stimuli this is called a "within subjects" experiment. If the participants are divided in groups that see different stimuli, it is called a "between subjects" experiment. The choice will affect the number of participants you need to achieve sufficient power (see Chapter 11) so you should use the setup that is most appropriate to your study.

The choice of the task that your participants will perform is crucial and is usually dictated by the methodology and/or by the prior studies that have manipulated the same variable, which you discovered in your literature review. The point here is that you should try to manipulate the variable you are interested in as efficaciously as possible and there is no need, usually, to reinvent the wheel.

Coming now to the specifics of eye tracking, there are all sorts of potential pitfalls unique to eye tracking, which would not be covered in a general

TABLE 10.1 2 × 2 factorial design

	Audio	No Audio
L1	L1 + audio	L1 – audio
L2	L2 + audio	L2 – audio

introduction to the experimental method. For example, each eye tracker has accuracy and precision measurements, that is, a measure of how close to the position of the actual stimulus the measurements are (see Figure 10.1).

A good way to visualize this is a dart player throwing darts at a bull's-eye: precision is how closely bunched the darts are; accuracy is how close to the center of the bull's-eye they are. When dealing with small stimuli (and in reading letters are usually fairly small) accuracy becomes crucial. So font size and the actual font used may be significant. In the example below (Figure 10.2), a relatively small error would translate into a hit or a miss on the AOIs on the eyes or mouth: note how the fixations (the dots) are outside the AOIs, in the image on the right, which has a vertical inaccuracy.

Orquin and Holmqvist (2018) reviewed more of the potential pitfalls of eye-tracking studies. It is essential that you familiarize yourself with the literature in your specific (sub)field, to be able to avoid these pitfalls by following received methodological practices. See also Carter and Luke (2020).

Usually, eye trackers come with software that allows you to build and run the entire experiment. The program should also allow you to display

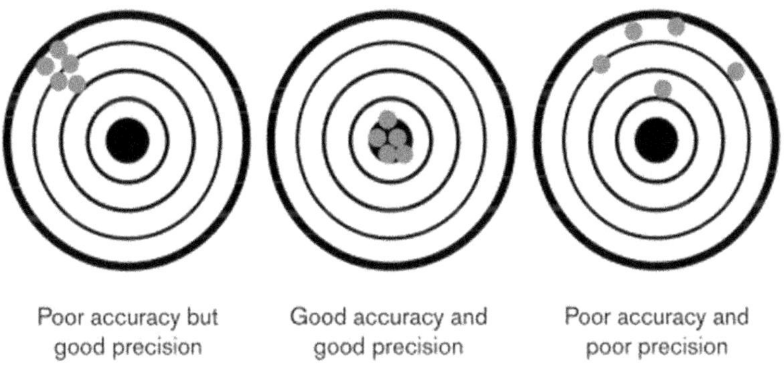

FIGURE 10.1 *Accuracy versus precision.* Source: *Tobii, reproduced with permission.*

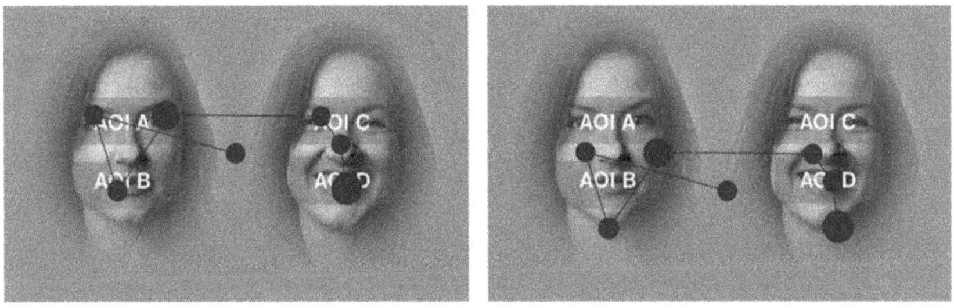

FIGURE 10.2 *An example of accuracy error. The figure to the left is accurate; the figure to the right displays a vertical accuracy error of less than 3°.* Source: *Tobii, reproduced with permission.*

slides on which the stimuli and/or instructions are displayed. Conklin et al. (2018) compared three such programs and there are many more. Each has its strengths and weaknesses, so you should determine which software to use depending on your needs and budget. There are also open-access programs that have some of the functionalities of these dedicated programs. Needless to say, psycholinguists and psychologists also have dedicated tools that can be used to deliver the experiment. Whichever tool you choose, make sure to test it thoroughly *before* you really need it. Ideally, you will have built and run a pilot study before having to do the real thing.

Finally, you probably will not have to worry about choosing an eye tracker: you will simply use one of those available to your department/advisor, and you will receive advice from them on whether the instruments that you have available are appropriate to do the research you want to do. This may determine right there and then whether your study is feasible or not (or you may need to get access to another machine that can handle the study you want to do). Should you be in a position of having to decide which eye tracker to purchase, Conklin et al. (2018, pp. 31–2) have a useful checklist to help make that decision.

Human Subjects Compliance and Ethical Treatment

The first step, which you should do even before you start planning your study, is to familiarize yourself with the human subjects' compliance and ethical treatment guidelines and obligations at your institution. The government, and the university with which you are affiliated,[7] want to make sure that you are treating your participants ethically and that you do not cause them harm.

Generally, these matters are handled by Institutional Review Boards (IRBs), Institutional Ethics Committees (IECs), Research Ethic Committees, or Ethical Review Boards.[8] While there are generally accepted guidelines (e.g., the World Medical Association Helsinki Declaration or, in the United States, the Belmont Report), each institution implements them differently. What may be a matter of course in one university may be unacceptable at another. Generally there will be institution-specific materials for you to study to understand what your responsibilities and requirements are. Also generally, universities in the United States require you to do some training, to ensure the proper treatment of your participants. Your responsibilities as an ethical scholar may range from protecting the rights and privacy of your participants and ensuring their safety and well-being to rather more trivial sounding details as to what refreshments if any you will offer them or where you will keep the data you collect and for how long. Be aware that different countries may follow quite different protocols. So, just because you are familiar with, say, the US protocols, that does not mean that Brazil, for example, follows the same rules. The National Institute of Allergy and Infectious Diseases (part of the National Institutes of Health) maintains a useful page comparing human subjects regulations across various countries (https://clinregs.niaid.nih.gov) as well as the International Compilation of Human Research Standards (https://www.hhs.gov/ohrp/international/compilation-human-research-standards/index.html). Another potential issue is compliance with the European General Data Protection Regulation (GDPR), which came into effect in 2018 and protects European Union citizens' personal data.

By knowing ahead of time what the requirements are that you will need to meet you will easily go through the various approval processes and you will avoid delays and problems. Generally, and again you should check your institution-specific procedures, you will need to file a proposal with an office or body of your institution in which you detail what you intend to do and how it affects your human subjects (if you have no human subjects, then you generally need not do so—but again, you should check). Your institution may require a detailed proposal, or a more schematic treatment, or perhaps they may require you to appear in front of an ethics committee. Whichever the process it is likely to take many weeks and sometimes months, especially if the ethics committee meets infrequently. Hence our suggestion to familiarize yourself with the procedure and take care of its requirements as early as possible in the process.

Since eye tracking requires video recording your proposal may attract extra scrutiny (after all you cannot cover the eyes of your subjects to protect their privacy!) and so you may need to have a more stringent protocol. Be aware also that projects with more than one principal investigator or even just with more than one researcher, or with researchers in various countries, may also require more extensive and/or more stringent criteria. Again, start early and familiarize yourself with the procedures and protocols of your institution or of the institution in which you are working/collecting the data.

Open Science and Data Sharing

Another topic which you should consider early on in the planning phase is whether you should share your data and/or preregister your hypotheses. Both these issues belong to a relatively recent trend in applied linguistics, but one which is showing great promise and that has significant repercussions for science at large, that is, the movement toward open science and data sharing. Open access, that is, the idea that one should be able to freely access scientific results, is obviously an attractive idea. This is not the place to go into the (valid) reasons for data sharing, but let us mention in passing the significant crisis of replications experiences in psychology (Wiggins & Christopherson, 2019). Replication, that is, the repeating of an experiment to test its results, is an underappreciated and often ignored part of science. However, it is crucial to establishing scientific results on a firm foundation. It is only when a given study has been replicated repeatedly, by independent researchers, that we can safely assume its results are solid. The key to replication is the sharing of the data sets used for the experiment/research. While not all replications involve using the same data sets, an important part of the scientific process is the scrutiny of data or their reuse. Another aspect of open science facilitated by repositories is the preregistration of hypotheses and methodologies. This prevents the practice of trying multiple hypotheses and statistical analyses until finding one that yields a significant result.

There exist many repositories and websites where data, hypotheses, and procedures can be shared/published. For example, the Center for Open Science (`https://osf.io`) or the IRIS database (`iris-database.org`), which is dedicated to applied linguistics (Marsden et al., 2017). Before registering your study and/or sharing your data, you should thoroughly research the policies of your institution, in particular your institutional protection of human subjects, as there may be concerns with publicly sharing data that have not been explicitly consented for that use. If you are writing a dissertation or thesis, you should consult your advisor prior to any sharing or registration, as there may be specific policies in place.

Performing Your Study

So, you recruited your participants and the time has come to actually run the experiment. Ideally, you will have some help (a fellow graduate student, or a helpful undergraduate, or your spouse, if they really like you), but in what follows we will assume a solo experiment.

Practical Tips on Data Collection

If possible, use a windowless room, for constant light conditions (especially for pupillometry studies), or collect the data after sunset using constant electrical illumination. Unless your experimental setup makes it problematic, seat the participants in a chair that cannot rotate and cannot be moved up and down by the user. Avoid anything distracting in the peripheral field of vision of the participant, particularly moving objects (e.g., screens), and silence all phones and tablets. Give the participant a practice task *before* starting the calibration process. This way they will "settle in" and any problematic behavior (such as leaning in toward the screen) will be apparent prior to the calibration process of the experimental phase.

Ask users to remove hats, especially with a brim, or baseball hats. If possible avoid heavy mascara (have some makeup-removing towelettes handy), eye glasses (unless essential), bangs over the eyes (provide barrettes), and drinks (which may cover the face). If possible, avoid having the participants touching their face or leaning face on the palm of the hand (especially in long tasks). You should offer the participants a drink and a small refreshment (e.g., cookies, fruit) after the session and don't forget to thank them for their time.

Have a protocol sheet that you or the person doing the recordings can refer to, with a check list (this will help in keeping track of the protocol). If possible, do a post-session debriefing (open-ended questions) but do not show the participants the recordings (e.g., scan paths) unless it is crucial to your protocol. Avoid "think-aloud" protocols: talking about an item will make the participant look at it, thus distorting the normal viewing (Nielsen & Pernice, 2010, p. 62).

Have a lab book in which you log the experiment. Make sure that the lab book and all the consent forms and other sensitive information are locked up when you are done.[9] Back up your data redundantly; that is, have more than one backup, in different locations. If possible one of your backups should be on the cloud.

Validity

It is important to validate your raw data (i.e., the measurements of the eye tracker). Usually, the eye tracker will provide you with a validity score for each sample it takes. Normally there will be a validity score for each eye. For example, on the Tobii X2 60 eye tracker that we use in our lab with the software provided (Tobii Studio), there are four possible scores, summed up in Table 10.2.

This may be confusing outside of the 0 and 4 scores, until one remembers that there are two eyes to find and the system does not know which one it has found. So a score of 1 is uncertain but with a preponderance of evidence for this eye and will be matched by a score of 3 in the *other* eye. A solution to avoid this complexity is to only accept as valid samples that scored 0 (i.e., of which the system is certain). On the one hand, this may cause you to lose a bunch of data: in a study we did we ended up discarding 13 percent of the recordings, which is painful, but on the other hand, you don't want unreliable data. The validity of the data is another reason to collect a larger sample than you will actually use in your study (see Chapter 11 on the issue of sample size).

TABLE 10.2 Validity score on a Tobii X2 60

0	The system is certain it has found both eyes
1	The system has found both eyes, and the one scored 1 is probably correct
2	The system has found one eye but does not know which one this is
3	The system has found one eye and thinks this is not the one
4	No eyes found

Handling Outliers and Missing Data

A special methodological problem faced by eye-tracking researchers is the handling of outliers and missing data. While these problems are not unique to eye tracking (they are common in all statistical processing) they become particularly significant in eye tracking due to some of the specifics of the technology. First and foremost, eye trackers all too often will lose track of the eye(s), for example, after blinking or after a head movement. This results in missing data that can sometimes be extensive. A way to handle this is to remove all trials with greater than 25 percent of track loss. Care must be given to avoid skewing the data, for example, by removing most of the trials of a participant. There are numerous statistical techniques for handling missing data, and depending on the circumstances of your study, it may be possible to interpolate the data or replace them (often a strategy is to replace the missing data with the mean figure or with the prior value, although these strategies are often decried), or use more advanced methods that use statistical procedures to estimate the missing data based on the observed ones (Enders, 2010). If the data are missing randomly, then they can be safely ignored. If the accuracy of eye tracker deteriorates over time (a phenomenon called "drift") recalibration may be necessary. Sometimes if the error is systematic, it can be corrected by moving the AOIs (e.g., suppose that you have an AOI consisting of words and your data show the fixations to be along the line of words but a few pixels underneath them). This sort of correction may be proper, depending on the circumstances, but always needs to be documented (Carter & Luke, 2020, p. 56).

Outliers are a common problem in statistical studies, because they can significantly skew the mean of your data: consider if you are assessing the income of 100 subjects and one of them is Elon Musk.[10] There are two types of outliers: errors and nonerror. Error outliers are typos when hand-coding information, for example. They can be simply corrected or deleted, if it is impossible to reconstruct what the real value should have been. Nonerror outliers are more interesting, as they are real data, just extreme values. Since they may significantly affect your results, the best approach is to check what happens to your data if you run your statistics with the outliers and without them. If the result changes, then you should report *both* results. If the result is the same, you can just remove the outliers, mention in your methodology that you did so, and carry on.

There is no universal criterion to identify an outlier, although for normally distributed data, 3 standard deviations from the mean (see Chapter 11) is a common criterion, so generally researchers resort to visual inspection of the data. For example, one might plot their data according to two variables and visually inspect the data for points that are very distant from where the rest of the data cluster. The problem in eye tracking is that if one is using scan paths, visual clutter is very likely to result from observing more than a few participants at a time. Various graphic methods have been developed to alleviate this problem, for example, by highlighting outliers using color (e.g., Gu et al., 2017) or other graphic methods. Here too, only prior experience of other scholars can be your guide, which is another reason to do your homework in the literature review.

Writing Up Your Study

You have completed your study and collected your data. You ran the statistical analysis with the help of a couple of friends and pizzas. Now you need to write up the paper. We will assume that you wrote the introduction, stating what the topic of your research is, the literature review, and the methodology, probably recycled from the description in your ethics review board application, ahead of time. What is left is often confusing for students who are not familiar with the organization of studies in psychology and the social sciences. The outcome of your study is presented while keeping separate the quantitative results of your study from their interpretation (i.e., the conclusions).

Describing the Methodology

You should provide as much information as practical about the physical setup of your experiment. The goal, as Carter and Luke (2020) point out, is to provide enough information so that someone may replicate your study. Thus you should provide the following information:

The model and manufacturer of the eye tracker used in the study;

The sample rate of the eye tracker in Hz (e.g. 120 Hz, 1,000 Hz);

The accuracy and precision of the eye tracker;

Whether one or both eyes were tracked. If only one eye was tracked, which one (left, right, dominant)?

What was the eye tracking setup (desktop mount, remote tracking, head-mounted)? Was a chin/head rest used?

The make and model of the monitor;

The size and resolution of the monitor;

Viewing distance from the eyes to the monitor;

The stimulus size(s).[11]

Finally, you should describe the setting in which the participants operated: sitting at a desk or in an armchair, and so on. Some reviewers or journals may object to some of this information and you can always take it out, but in this case it is always better to err by excess.

Results versus Conclusions

The idea is not to contaminate the objective nature of the results from their subjective interpretation.[12] So, you should first present the results that were recorded and then, in a separate section, what the results mean, that is, have your hypotheses been falsified?

Writing Up Your Conclusions

The conclusions are thus a broader discussion, usually encompassing all the results you established in your study. Usually one will have four or five hypotheses, and these will not be independent of each other; in other words they will be mutually supportive. So, again speaking generally, your discussion will recapitulate the findings about the various hypotheses and then attempt to synthesize an overall conclusion, a broader general point that is supported by the totality of your work.

Usually, the conclusions are followed by a short discussion of the limitations of the study. This is where you will note any unexpected issues that arose during the study (e.g., you had to discard more subjects than expected due to calibration issues). Here too you should discuss the boundaries that were more or less implicit in the **operationalizations** of the concepts you performed at the beginning of the process. For example, you may note that your study of gaze in NS/NNS interactions is limited to dyads or that the experimental setting was not precisely conducive to a friendly conversation. You should be candid about these issues, but you should remember that science is not done on the basis of intentions, but with results. If the problems with calibration caused your sample size to be too small for the statistical power of your experiment to be adequate, it's too bad but your experiment is underpowered. Don't panic, there are ways to palliate the issue; for example, the power may be enough for some of your hypotheses but not for all of them, because, for example, the effect size to be detected is larger in some of them. Alternatively, you might be able to run a different statistical test, with lower power requirements.

Conclusion

Setting up your study and your research methodology is crucial to having a successful experience doing eye-tracking research. The literature review contains the collective wisdom of generations of scholars who have found best practices and ways to handle the problems and issues that arise in doing research on the topic of your choice. While keeping a critical eye on the received wisdom, you should take full advantage of that knowledge and that of your supervisor or mentor. Remember our admonishment about the solitary genius myth being increasingly rare in eye-tracking studies. Your dissertation or thesis may be the only times that you will single author a piece of writing, unless it is a survey or meta-analysis.

Be sure to develop your methodology, including the statistical treatment, prior to submitting your application to the responsible oversight committee for human subjects research, partly not only because you will most likely be asked to detail it in your application but also because planning your study in some detail is indispensable for success.

Self-Study Questions

The following questions pertain to the contents of this chapter. You should be able to answer them without looking up the answer in the book.

1. Are the procedures to handle human subjects the same in every country?
2. A student comes to you with an idea for a study. They have not read the literature because they don't want to taint the originality of their idea. What do you tell them?

Words to Know

literature review	methodology research question operationalization
gap in the literature	

Further Reading and Sources

The Belmont report can be found here: `https://www.hhs.gov/ohrp/regulations-and-policy` belmont-report/index.html. The World Medical Association's Declaration of Helsinki can be found here: `https://www.wma.net/policies-post/wma-declaration-of-he`.

CHAPTER 11

Principles of Statistical Analysis

Why Statistical Analysis?

In eye tracking, statistics is practically speaking indispensable. So you will need to master some statistical knowledge. In what follows we will outline the basic concepts that this entails. Let us be very clear: what follows is not and could not be a full introduction to statistics. That would require a book in itself. What this chapter aims to do is describe how to set up your research plan in such a way that you will be happy when you do the statistical analysis. The process goes like this: you need to know a few things about statistical analyses in general to decide what kind of experiments to run, what **variables** to set up, how many participants you need, and so on. Once you have done all that, you will have all the information needed to select the statistical test(s) that will give you the answers you need. Our goal in this chapter is to guide you until the selection of the test. It would be impossible in this book to go into the detail you need to actually run the test. Statistics is an academic field, just like linguistics or psychology: every test has a vast literature about it and you will need some understanding of it, especially if you are not using one of the most common tests. When you are researching the test, the foundational concepts we will discuss will help you understand the more advanced discussion, or if you have the opportunity to work with a statistical consultant (and we have found that statisticians are generally quite happy to help), these very same concepts will help you understand what the statistician is telling you. For some of our readers, the mere talk of statistics carries a certain amount of trepidation. However, in this presentation we have avoided the traditional mathematic approach to explanation, focusing instead on the concepts and avoiding formulas, as much as possible.

What Kind of Questions Do You Have?

The first question you should ask yourself is, what kind of questions am I asking? For example, do you want to know what your data look like: for example, is this set of behaviors larger than this other set, and if so could the difference be due to chance? Or do you want to know, given a sample of data, how will this generalize to all the data? This is an important first step, because the way the data are collected will differ depending on the question. The first type of question, directed at figuring out what the data look like, or summarizing them, is called **descriptive statistics**, whereas the second type of question is called **inferential statistics**.

In the case of descriptive statistics, we take the data we have pretty much as they are. The purpose is to understand them, to summarize their characteristics, and to make it easier to grasp them. Suppose you have the grades of the students in one class and you are interested in how the class is doing (i.e., did the class as a whole pass the test?). The data will probably come to you as a list: on one side the name of student and on the other side the grade they scored (A, B, C, etc., or 100 percent, 95 percent, etc.). Suppose now that the class is fairly large and you have 150 students. Reading through the list is going to be time-consuming and difficult: it will be hard to remember who passed and who didn't. However, if you calculate the average score, that one figure will tell you *overall* how the class is doing (if they scored 83 percent, e.g., the class is probably doing quite well). So the average can be thought of as a summary of the scores it averages over. Now, much like a summary of *Romeo and Juliet* as "an unhappy love story" is missing a lot of what makes *Romeo and Juliet* worth reading, summarizing data means setting aside some of the information. For example, knowing that the average score was 83 percent is unlikely to be much comfort to the student who scored 40 percent and is now failing. However, we need to remember that our question was "how is the class doing?" not "how are individual students doing?" In other words, there is a trade-off between summarizing data and the complexity and detail of the data themselves.

Let's now turn to inferential statistics. Our goal here is to generalize, from our **sample** to a **population**. In other words, we are not just interested in our data (say, the 150 students in the class); we want to be able to make claims about *all* students. Here you may object that making a claim about "all" students is an impossible task. Indeed, it is. A population is generally defined as a well-defined set of "entities" (things, people, etc.). What that means is that we define the scope of our inquiry. When we say "all students" we likely mean all the students in this country, or school, or class, as of today or some other point in time. The point is that the definition must be as clear as possible, because if it's not clear who or what belongs in the population, then nothing of what we may say about it is really meaningful. Usually, defining the population, at least as far as applied linguistics goes, is unlikely to be a problem. You will use definitions and criteria from applied linguistics. For example, let's say you want to compare native speakers to Lx speakers. The definitions will come from applied linguistics but you will probably need to operationalize them anyway. For example, people do not come with a label saying "native speaker." So, you will need to ask them if they grew up speaking a given language. That is *one* way of operationalizing the concept. Needless to say, there could be others. The distinction here is between a theoretical definition (the kind you find in textbooks) and an operational one, which is the one you work with.

However, having a well-defined population is not enough. In order for inferential statistics to be valid we have to make sure that the sample we are working from is representative of the population to which we want to generalize. This makes a lot of sense but is harder to do than one would think. There is a process that needs to be followed carefully. In inferential statistics, you start with a **hypothesis**, an explanation of the way you think things are: for example, this method of teaching improves the performance of students. That's a hypothesis (essentially, an informed guess as to the nature of the world). In inferential statistics we test the hypothesis and, based on the results, we either accept or reject the hypothesis. The acceptance or rejection of the hypothesis is done on the basis of the p-value of the statistics we are using (more on that later in the text).

The **null hypothesis** is essentially the idea that there are no differences or correlations in your data. Suppose that you measure the performance of two groups of students. The null hypothesis is that there is no difference in performance between the two. That does not necessarily mean that they perform the same, only that the difference between the performance of the two groups is not significant. Suppose that you tested 100 items and student *a* scored 99 percent correct, whereas student *b* scored 98 percent correct. The difference (1 percent) is not significant in most contexts. For example, both students would get an A grade on the test.

Sampling

Suppose that you want to know what people in the United States think of the band Insane Clown Posse. You think about how to go about gathering data. The obvious thing to do is to ask all the inhabitants of the United States. This presents an obvious drawback: there are a lot of people in the United States (about 300 million) and asking each and every one of them is, practically speaking, impossible.[1] Since asking everyone is not possible, you need a sample and, since not everyone knows about the band, you figure that you might go to one of their concerts and interview the people there. However, when you think about it, that's not a good idea, because only people who tend to like a band will pay to go see it perform.

What to do? Fortunately, statistics has a whole field dedicated to **sampling.** Sampling is the act of selecting a sample from a given population to which you wish to generalize such that the sample is representative of the population. What that means is that every person or sampled item has to have the same probability of being included. That is why going to the concert was a bad sample: the likelihood of it containing many people appreciative of the particular type of hip-hop associated with Insane Clown Posse is pretty high. But if you ask random people in the street, most of them will not have heard of the band. So, you would need to collect a very large sample of the population of the United States, if you wanted to generalize to the whole country. Or you could subdivide your population in two halves: those who know about Insane Clown Posse and those who do not and collect data accordingly. This is called stratified sampling. Outside of the facetious example, stratified sampling is particularly useful when you want to make sure that enough members of a target group are included. For example, if you are interested in how low-income students, in particular, among students in general, deal with the cost of textbooks, you should make sure that through stratified sampling you have enough low-income students in your sample for your generalizations to be meaningful.

Variables

The second question you should ask yourself is "what are my variables?" This may sound very complicated, but in fact it just means "what are my questions?" or "what kind of data do I have?" Let us consider again that question above: we want to know how students cope with the cost of textbooks. Suppose we wrote a questionnaire and distributed it among a representative sample of students, stratified in such a way that we have a good-sized group of responses from low-income students. Obviously, "income" is the first variable we have to deal with, since we wanted to know how low-income students in particular dealt with the problem. There are numerous ways to operationalize this for our study: we could check how many of them are on financial aid, or we could ask them what their annual income is. There are different ways of asking this question: one is to let them fill out a box and the other one is to give them ranges, for example, less than $25,000, $25,000 to $50,000, and so on. Note that in the first case, we have a yes/no bit of information: is so-and-so on financial aid, yes or no? In the second case, we have a figure, an actual dollar amount. In the third case, we have a range: someone may be in the first group or the second, and so on.

Other variables in the study would be what they think of the issue (it's a major problem, a small problem, not a problem, not a problem at all), how much they spend on textbooks per course, whether they buy the books or not, their major, and so on. Each one of these is a variable. Each one of these will be a different set of data, depending on how it's been operationalized and the data have been collected. They could be yes/no questions, exact figures, ranges, scales, or, in the case of the major, just the name of the major (English, Linguistics, Statistics, Chemistry, etc.).

The reason this is important is that *depending on the type of variable you have, the type of statistical analysis is likely to change*. In other words, there isn't a one-size-fits-all statistical technique that works for every type of questions or variable. The type of variables you have determines the choice of statistical test you will use. In the rest of this section we will walk you through the process.

Taxonomy of Variable Types

So, what kind of data are there?

- An important kind of data are **categorical, discrete, or nominal data**. Examples of categorical data are nationality (US, Chinese, Indian, etc.), car make (Chevrolet, Ford, Toyota, etc.), geographical location, hair color, blood type, and so on. There is no intrinsic ordering (should Ford come before or after Toyota?). Note that alphabetical order is a pure convention. There is no good linguistic reason for the letter "a" to come before "e." In fact, in English "e" is more frequent. The most important thing about categorical or nominal data is that there is no quantitative relationship between the values of the data. Consider the case of the major, in the example of the textbook cost survey. There is no order or quantitative relationship between English, Computer Science, and History.
- A special kind of categorical data are **binary data** (or dichotomous, from the Greek "cut in two"). They are things like true/false, yes/no, pass/fail, married/unmarried. In our example above: receives financial aid is a binary variable.
- Another kind of data are **ordinal data**, which are expressed in numbers but exist only as a ranking such as first, second, third or the days of the week and other ordered or ranked sets. Examples are first fixation in eye tracking or the top three students of a class. There is no assumption that, say, the second-ranked student in a class did half as well as the first ranked. Neither categorical nor ordinal data are quantitative. In our textbook example, we might have asked the respondent to rank the most expensive classes, second most expensive, and so on.
- **Quantitative data**. Most of the data you will be working with will result from counting things (e.g., number of students enrolled in a program, participants in an experiment, etc.) or measuring them (volume, pitch, length of pauses, etc.). These are all quantitative data

TABLE 11.1 Types of data

Categorical	Discrete, Nominal
Binary	Dichotomous
Ordinal	
Quantitative	Continuous, scalar

(Table 11.1). In our example, the dollar figures were all quantitative data.

Quantitative data are **scalar** (i.e., they can be naturally ordered). Statisticians distinguish between *interval scales and ratio scales*. A good example of interval scales are temperatures. Most Likert-like scales are also interval scales. Our example of whether the students considered textbook cost a major or minor problem is an interval scale. Ratio scales are so called because the ratios between numbers are constant: the difference between 50 to 100 and between 150 to 200 remains the same, unlike say in temperatures: 80 degrees is not twice as hot as 40 degrees. They are also called "**continuous**" data. Also, ratio scales are the only ones to have "meaningful" zeroes. This is a subtle distinction. Consider counting the number of people who showed up at your birthday party. If ten people show up you have twice as many as if five people show up. This tells us that it is probably a ratio scale. But also consider that if zero people showed up, literally no one came. Now contrast this with temperatures. Unquestionably the Celsius scale is far more rational than the Fahrenheit scale, since water freezes at 0 Celsius and boils at 100 Celsius. However, despite it being far less motivated, the Fahrenheit scale shows that zero is not a particularly meaningful point, in temperature: the fact that water freezes there is just a matter of convention. Indeed in the Fahrenheit scale water freezes at 32 degrees and boils at 212 degrees F.

You may encounter a different classification of variables as **dependent** and **independent** or **explanatory**.[2] The independent variable is the one manipulated by the researcher (e.g., native speaker vs. non-native), and the dependent variables are typically measurements (proficiency, intelligibility, scores in class, etc.). The dependent variables are so called because they depend on the independent variable which "causes" them.[3] If you compare reading times (dependent variable) in two groups (NS and NNS; independent variable) you are trying to show that it is the difference between native and non-native speakers that causes the differences in reading times.[4] You may ask in what sense did the researcher "manipulate" the NS/NNS variable.

The manipulation consists in picking two groups of participants on that basis. In other words, the researcher constructed the sample to be just so.

Another term that you may encounter is a **blocking** variable. Again, the terminology may be misleading. A blocking variable is just a variable that allows the researcher to divide the data into "blocks," for example, male versus female or students from class a, class b, and so on. Blocking variables are used in some statistical analyses to "eliminate" variables that are not interesting or relevant to the study. For example, suppose you collected data from your school and so you have data from five classes. You would use a blocking variable to eliminate the fact that the students come from these separate "groups." Obviously, if you wanted to study the differences in performance between the five classes, that is no longer a blocking variable, but it becomes your independent variable. It all depends on what you are trying to do.

Finally we have the dreaded **confounding variables**. A confounding variable is a "hidden" variable that has an effect on the dependent variables but that you are not aware of. For example, in scholastic testing, performance (dependent variable) was usually seen as the outcome of effort or intelligence (independent variables). However, it turns out that social class was a confounding variable: in other words, rich children outperformed poorer children not because they studied harder or were smarter but because they had better resources (access to and familiarity with books, tutors, parents able to invest time in their performance, etc.).

To put it simply: confounding variables are bad. You should do your best to avoid them. There are research design techniques to rule out confounding variables and once you have identified them to eliminate them. They are too advanced for us to deal with in any detail, but randomized trials are one way of reducing confounding variables, restricting the sample to subjects/items that are controlled for the possible confounding factor (e.g., using only monolingual speakers, to avoid influences from other languages), and matching the samples (e.g., making sure you have 50 percent NS and 50 percent NNS).

Testing for Normality

Once you have collected your data, the first thing you need to do is find out how they are distributed. Suppose you want to see the distribution of the grades on the final paper of your class. You take the papers and put them into piles: all the As, the Bs, and so on. All you need to do now is to count how many papers you have in the A pile, the B pile, and so on. One way to visualize this is to use a bar plot. The piles are your bars. Suppose, for example, that you had the following results: 14 A, 12 B, 6 C, 5 D, and 2 F.

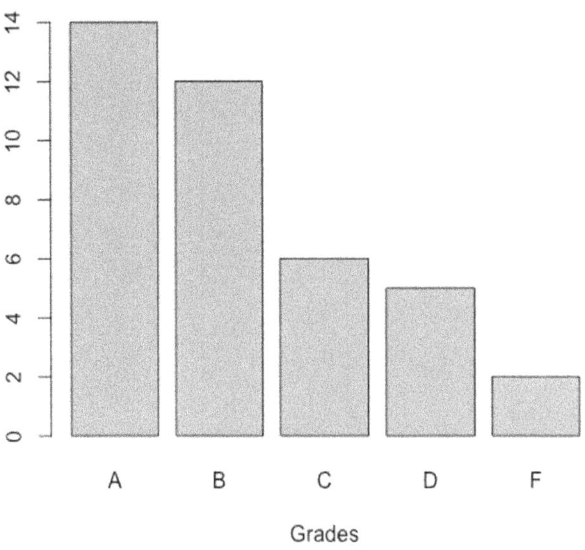

FIGURE 11.1 *A simple bar plot generated with R.*

On the horizontal axis (aka, the x-axis), we have the grades A, B, C, and so on. On the vertical axis (aka, the y-axis) the length of the bar tells us how many instances of that grade we had: for example, fourteen students got an A and the A bar is as high as 14 on the y-axis.

Histograms

Histograms are a little more complex but very useful; for example, suppose you have a large class of 150 students. You give them a first exam and you want to see how the class did. Counting the various scores would be tedious and time-consuming. One way to do so is to put them into "bins." The process is very easy because R does the binning for you automatically (you can also specify it by hand, if you prefer). A **histogram** is a visual representation of the bins. We generated a random sample of 150 "scores" between 1 and 100, as a simulation of the scores of an exam with 150 students. This yields the following histogram: on the x-axis we have as in the bar graph above the distribution of the grades, in bins (i.e., 0 to 10, 10 to 20, 20 to 30, etc.)[5] and on the y-axis we have, again as above, the number of students who scored those grades. So, from the graph we can deduce that thirteen students would have gotten 0 to 10 out of 100.

Histograms are meant to visualize the distribution of a continuous variable, such as scores or, say, the height of people. If you have a discrete variable, such as first-year student, second-year student, and so on, you would use a bar plot. The purpose is the same: to visualize what the distribution of

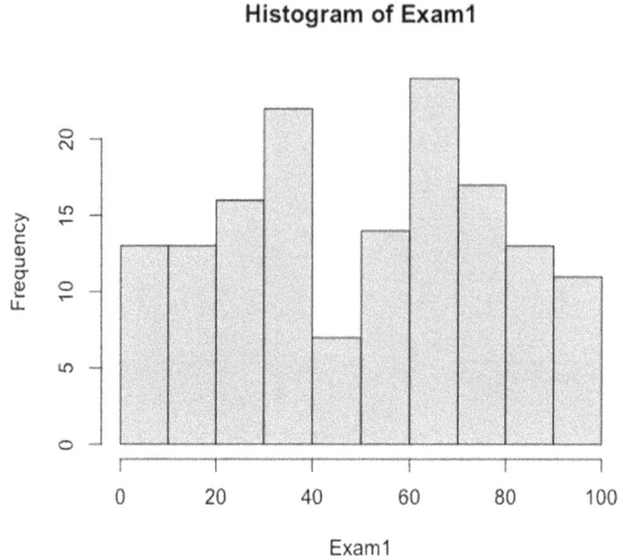

FIGURE 11.2 *A simple histogram generated with R.*

the data looks like. A visual representation is often more intuitively grasped than a list of numbers.

Histograms have another useful aspect; they can be used to "eyeball" whether the distribution of the data is normal or not, by checking visually if they represent a **bell curve**, or not. If you compare the histogram in Figure 11.2, which is a non-normal distribution (they are random numbers), to the one in Figure 11.3 you should notice that the histograms have very different shapes. Compare the histograms above to the one below, in Figure 11.4. The histogram in Figure 11.3 resembles the one in Figure 11.4.

Normality and the Bell Curve

The bell curve, so named because it recalls the shape of a bell, represents a **normal distribution**. The word "normal" is here used in a mathematical sense. The fundamental property of the bell curve or the normal distribution is that it is symmetrical around the mean (the dashed line in Figure 11.3). In this case, the mean is 0.

Most of the data points occur in the taller central area, whereas only a smaller fraction occur in the tails (i.e., the areas where the end of the line is almost flat). In fact, we can be more precise by introducing the concept of **standard deviation.**

Standard Deviation. We saw that a bell curve is roughly centered around the mean of the data. Recall that the mean (or average) is equal to the sum of the data, divided by the number of data. For example, if we have three

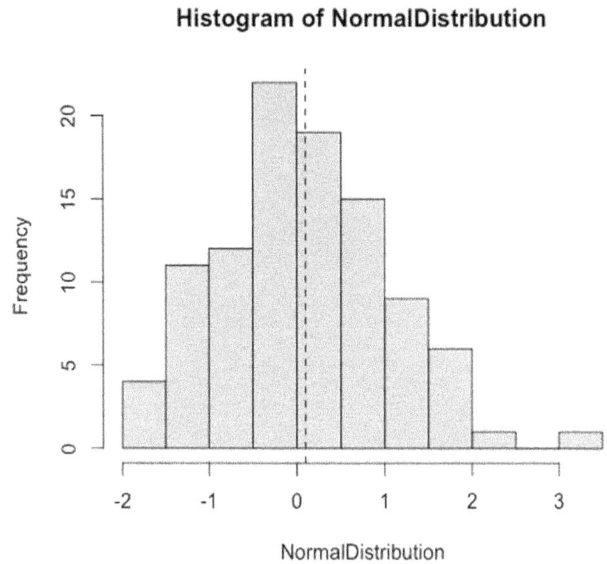

FIGURE 11.3 *Histogram of a random sample of normally distributed numbers.*

scores in a test of 90, 100, and 82, the mean is calculated by the formula: $\frac{90+100+82}{3}=\frac{272}{3}=90.66$

Intuitively, the standard deviation is on average how far from the mean the values of the data "stray." Mathematically, we need to define the concept of **variance**. Variance is simply the deviation from the mean. For example, suppose that the mean score of an exam is 80 points. If someone scored a 92, the variance of their score is +12. If another student scored 76, their variance is −4. To get rid of the negative numbers, statisticians simply square the numbers (the square of a negative number is a positive number). If you then take the mean of the squared distances from the mean you get the variance of the sample.[6] Finally, if we take the square root of the variance, we get the standard deviation.

The size of the standard deviation will affect the shape of the bell curve, as can be seen in Figure 11.5. A smaller standard deviation will mean that the data are gathered closer to the mean (here set to 0 in all three cases), whereas a larger standard deviation will mean that the data are more scattered around the mean.

An interesting detail is that about 95 percent of all the data lie within two standard deviations from the mean, so a good way to think of the 0.05 value of the level of significance (**alpha**, see Alpha, aka, Significance Level section) is that it draws the boundary between four standard deviations (two on each side of the mean) and the "tails" of the distribution. In Figure 11.6, the tails

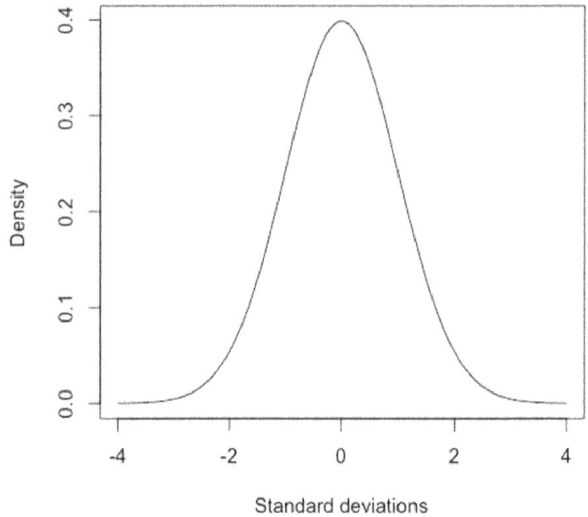

FIGURE 11.4 *Bell curve (normal distribution).*

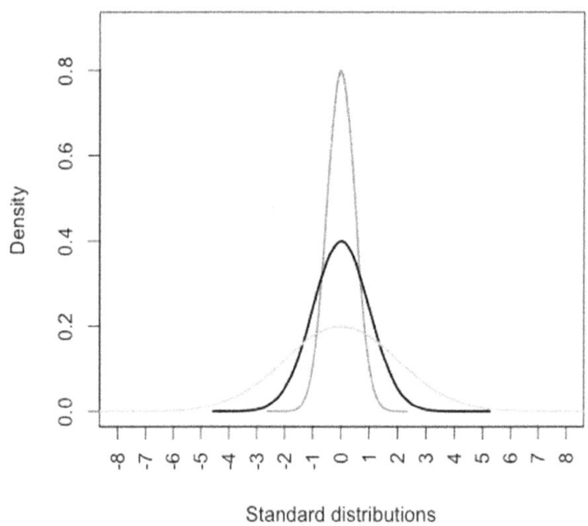

FIGURE 11.5 *Normal distributions with different standard deviations.*

(shaded area) are to the left and right of the (roughly) −2 and 2 standard deviations points, respectively. This visual is often used to represent alpha.

Now that we have discussed normal distribution, we can explain one of the reasons that linguists without a mathematical background often get

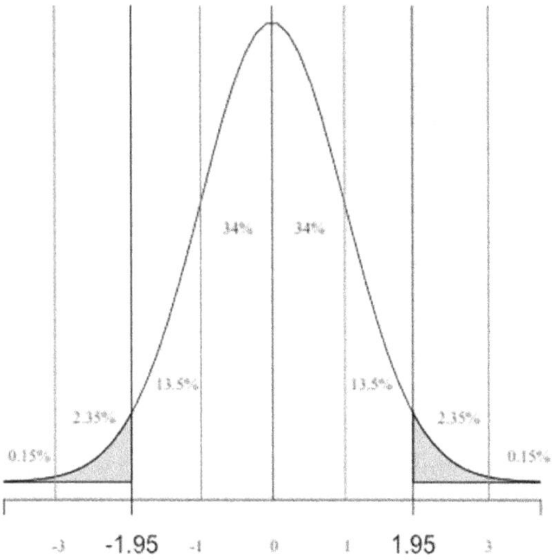

FIGURE 11.6 *Normal distribution with standard deviations and percentages of data. The mean is set at zero (central line).*

frustrated with statistics. Most traditional treatments of statistics start with normal distribution and dedicate a very significant amount of time and energy to discussing it. However, few linguistic data are normally distributed. This can be very frustrating for a linguist taking a statistics class (usually offered in the psychology department). This is not to say that knowing about normal distribution is not useful, of course. It is. But linguists cannot stop here and need to go further afield. We will return to the differences between general statistical methods and those used more frequently in linguistics.

Choosing the Statistical Test

Once you know whether your data are normally distributed or not, you can start the process of determining which statistical test is appropriate for your study. The choice of test depends on a number of factors examined in what follows.

Parametric versus Nonparametric Tests

The reason the question of whether your data are normally distributed or not is important is that it affects the choice of statistical tool you should use. Specifically, if your data are normally distributed you should use **parametric tests.** If they are not, you cannot. This is because parametric tests assume the data are normally distributed. Nonparametric tests do not. Generally speaking, for whatever test you need there will be a parametric (i.e., assuming normal distribution) and a nonparametric test available (i.e., not assuming normal distribution).

There are useful charts that guide you in the choice of test, for example, the one shown in Figure 11.7, from Campbell and Shantikumar (2016), is quite comprehensive. Recall that an input variable is the independent variable (the one manipulated by the researcher) and the outcome variable is the dependent variable. It should also be noted, as Campbell and Shantikumar (2016) noted as well, that these guidelines are merely indicative and that each case should be analyzed independently. This is when a consultation with an expert may be advisable. Some universities have programs, run out of the statistics department, to help other scholars on campus, and sometimes research libraries have similar programs. Consulting with experts in the field may also be an option. As a last resort one may actually try and study statistics (we know, we know, but bear with us anyway). Comprehensive treatments of statistical analysis exist for each test.

However, this does not mean that nonparametric tests do not make other assumptions. You should always thoroughly research what assumptions a test makes *before* using it, so as to make sure that you are using the right test for the data at hand.

Independence

The assumptions that a test makes will vary. For example, many tests assume that the observations are independent, that is, that each data point that you collect does not affect any other data point. For example, if you test forty students randomly selected from the set of all students enrolled in the university on their capacity to identify passive sentences, the capacity of student *a* will likely have nothing to do with the capacity of student *b*, and so on. Conversely, suppose that you selected forty students from the class that you happen to be teaching. You cannot rule out that there may be students who are enrolled in another class who would perform differently. For example, your reputation as a stern disciplinarian obsessed with the passive may have discouraged weaker students in grammar from enrolling in it. So, while your observations of the capacity of randomly selected students to identify passive sentences data would be independent, the observations on the same skill set in your class would not. So, assuming that your data are normally distributed, and that the variables are quantitative, in the randomly selected case you would be able to use a t-test to compare students who received a treatment to a control group, but you could not in the other case.

In linguistics, a lot of conversational data are of necessity nonindependent: in a conversation between two people, if speaker *a* talks a lot, speaker *b* will

		Outcome variable					
		Nominal	Categorical (>2 Categories)	Ordinal	Quantitative Discrete	Quantitative Non-Normal	Quantitative Normal
Input Variable	Nominal	X^2 or Fisher's	X^2	X^2-trend or Mann - Whitney	Mann-Whitney	Mann-Whitney or log-rank[a]	Student's *t* test
	Categorical (>2 categories)	X^2	X^2	Kruskal-Wallis[b]	Kruskal-Wallis[b]	Kruskal-Wallis[b]	Analysis of variance[c]
	Ordinal (Ordered categories)	X^2-trend or Mann - Whitney	[e]	Spearman rank	Spearman rank	Spearman rank	Spearman rank or linear regression[d]
	Quantitative Discrete	Logistic regression	[e]	[e]	Spearman rank	Spearman rank	Spearman rank or linear regression[d]
	Quantitative non-Normal	Logistic regression	[e]	[e]	[e]	Plot data and Pearson or Spearman rank	Plot data and Pearson or Spearman rank and linear regression
	Quantitative Normal	Logistic regression	[e]	[e]	[e]	Linear regression[d]	Pearson and linear regression

[a] If data are censored. [b] The Kruskal-Wallis test is used for comparing ordinal or non-Normal variables for more than two groups, and is a generalisation of the Mann-Whitney U test. [c] Analysis of variance is a general technique, and one version (one way analysis of variance) is used to compare Normally distributed variables for more than two groups, and is the parametric equivalent of the Kruskal-Wallistest. [d] If the outcome variable is the dependent variable, then provided the residuals (the differences between the observed values and the predicted responses from regression) are plausibly Normally distributed, then the distribution of the independent variable is not important. [e] There are a number of more advanced techniques, such as Poisson regression, for dealing with these situations. However, they require certain assumptions and it is often easier to either dichotomise the outcome variable or treat it as continuous.

FIGURE 11.7 *Statistical tests for independent data. Source: Campbell and Shantikumar (2016), reproduced with permission of* Health Knowledge.

have a harder time talking than if speaker *b* had been paired with a taciturn speaker *c*. In cases like this, the assumption of independence is not respected, so you will need to find another statistical test. Another situation where obviously the data are not independent is when you test someone twice, for example, before and after you have taught them something to see if they learned it. Obviously there is no way that the second observation can be independent of the first one since it's the same person responding! In this case there are specific tests designed to handle paired data (i.e., when you survey a subject twice). Table 11.2.

Finally, it bears repeating that you should always check out what *all* the assumptions of a test are, because some of them can be quite specific. For example, the Chi-squared (X^2) test requires you to have a minimum of five observations per condition.

TABLE 11.2 Test for paired data

Type of Data	Test
Nominal	McNemar's test
Ordinal	Wilcoxon
Quantitative (non-normal)	Wilcoxon
Quantitative (normal)	Paired t-test

The Purpose of the Analysis

This is in a sense the most important choice you need to make: what is the purpose of the statistical test? Are you trying to determine if a difference in the performance of two groups is significant, or more generally are you comparing two samples? Or are you trying to assess whether two variables in your data are correlated? These are broad purposes. Other goals may be more specific, such as testing the degree of agreement between raters (e.g., you have several faculty members grading tests and you want to check that they are assessing the students roughly the same way). Or you may wish to assess the validity of the questions in a classroom objective test ("multiple-choice test"). Each of these cases requires different types of statistics, each with its set of assumptions.

Usually, within eye tracking, the most commonly used approach is linear modeling, that is, checking if there is any linear correlation among the variables in the study. However, the truth is that almost each study has its own methodology. As we discussed in Chapter 10, one of the reasons for doing a literature review is to find out what are the accepted, standard type of analyses (including statistical methods) for a given kind of study.

Wrapping Up the Choice of Test

We can thus conclude this section on choosing a statistical test to match our data by summing up the previous sections: the choice of test is determined by the following factors:

1. The normality of the data
2. The type of variables (e.g., quantitative, ordinal, categorical, etc.)
3. The assumptions of the test (e.g., independence)
4. The purpose of the analysis (description, correlation, regression. etc.)

It is very important to use the right test that fits the data and the variables. Indeed, as we saw in Chapter 10, the way the data are collected, questionnaires written, and so on may have to be planned specifically with the type of statistical analysis in mind.

The Quartet of Doom

How good of an answer to your questions your data are is determined by the interplay of four measures, which we have labeled for their significance and relative complexity, the quartet of doom. By the time we are done with them, the four horsemen of the apocalypse will be four playful kittens or cute puppies, depending on your preference.

Size of the Sample

The size of the sample is traditionally labeled *n*. This could be human subjects, words in a corpus, trees in a farm, or widgets from a production line. So, when you read $n = 15$, that means that the test was conducted on fifteen participants, widgets, or whatever was counted. Here you need to be cautious, because it is not always clear if the researcher means that they started with fifteen participants, for example, or whether after they eliminated all the participants that for whatever reason were not included in the test, they ended up with fifteen participants (so they may have started with twenty or even more participants). The biggest question in planning a statistical study is "How many participants do I need?" The answer is, of course, "it depends." More specifically, it depends on the following three factors: alpha (significance), **beta/power**, and **effect size.**

However, before we get to significance and power, we need to understand error. A large part of statistics is avoiding errors.

Types of Error

Suppose you are testing a given hypothesis. You can either accept or reject the hypothesis (H). Regardless of your prediction, H can be true or false. For example, suppose you invited a friend for dinner. Will she show up? You may think that she will or she won't, and she may show up or not. Your hypothesis is, say, that she will. So if she does come, H is true; if she doesn't, H is false. So, you can be wrong in two ways: (1) you could be wrong because you predicted that she would come and she did not or (2) you predicted that she would not come and she did.

TABLE 11.3 Types or error

	H false	H true
Reject H	Correct	Error (type 1)
Accept H	Error (type 2)	Correct

TABLE 11.4 Error terminology

Type 1	false positive	alpha	0.05
Type 2	false negative	beta	80%

This can be represented as a 2 × 2 table, like Table 11.3 below:

If the hypothesis is false we should reject it (top left box). If the hypothesis is true we should accept it (bottom right box). If we do, we are correct. If we do not, we incur in error. Rejecting a true hypothesis is a **type 1 error** (false positive). Accepting a false hypothesis is a **type 2 error** (false negative). Consider this example: at airport security if your keys make the security go off, it's a false positive (they mistakenly believed there was a weapon = positive). If they miss the gun you hid in your underwear,[7] it's a false negative (they did not catch the weapon). Type 1 error is called "alpha," whereas type 2 error is called "beta."

Alpha isusually set at 0.05, sometimes at 0.01. Beta is also conventionally set at 80 percent. These are arbitrary thresholds. What it means is that in five cases out of 100 we may reject a hypothesis that is in fact true. That's fine if we are investigating students' performance on the TOEFL, when your worst-case scenario is that you admit a few students who should not have been admitted. But in other situations, we want a possibility of error to be much smaller. Would you get on a plane that has *only* a 5 percent chance of falling out of the sky? Obviously not. So, the threshold of significance has to be much higher in that case. The general rule is that the smaller the alpha, the better. So, why don't we set our significance for scientific research much higher? In airline flights the number of accidents per flights is extremely low: in 2017 it was less than 0.5 accident per million flights,[8] or 0.000005. Surely, if we expect that level of safety in flying, shouldn't we want it in scientific research?

The problem is that alpha (significance) and beta are like two faces of a coin. If you raise alpha (i.e., accept more chance of rejecting a true hypothesis) you get a lower beta. For obvious reasons you want the likelihood of a false positive to be small (you don't want to claim you discovered something when you didn't). Moreover there are pragmatic reasons as well: your paper will not be published or your dissertation approved if you don't have a low enough level of significance. So, you might think that you should set your alpha to a very small possibility of being wrong (say, 0.001) but the effect would be to have very little power, since beta is the mirror image of alpha. We will discuss alpha and beta in more detail below. Table 11.4.

Alpha, aka, Significance Level

Significance (alpha) is, as we saw, the likelihood of accepting as true a false hypothesis. To put it differently is the likelihood that the difference between two samples or the correlation between two factors is due to chance and not to the way the samples are. For example, if you pull out a ball from a jar that contains ninety-five yellow balls and five red ones, the likelihood of pulling out a red ball is 0.05. So it is safe to hypothesize that the ball coming out of the jar will be yellow. Suppose now that you pull out a red ball: that's due to chance. The rest of the jar is 95 percent yellow. There is a consensus that for a result to be significant (i.e., be unlikely to be due to chance) 95 percent probability of being correct is good enough (sometimes you will see 99 percent).

The significance level determines whether to reject or to accept the hypothesis being tested in inferential statistics. Many statistical tests will yield a **p-value**,[9] which is the likelihood that one would get a result like the one in the data if the null hypothesis were true. Other tests have other similar values (e.g., r value or k); we need not worry about these in this context. Suppose, for example, that we get a p-value of 0.00001, that is, there is a chance in 10,000 to get a result such as the one we got. Hence, since we have decided to set our significance level at 0.05 and the p-value is *smaller* than that, we have to reject the null hypothesis, because it is highly unlikely that it is true. But if we reject the null hypothesis, it follows that the alternative hypothesis, the one we formulated, must be true! It is a little counterintuitive, since instead of trying to prove our hypothesis right we prove its opposite wrong, but once you get used to it, it actually makes good sense: recall that per Popper's principle of falsifiability no hypothesis can ever be proven correct, but any hypothesis can be falsified (i.e., proven wrong).

For a very long time, significance was the almost exclusive focus of the uses of statistics in the social sciences (and what little there was of it in the humanities). However, in the past ten to fifteen years, the focus has switched to include power as well. If one looks at statistics manuals from the 1980s, there is little if any mention of power and extensive discussions of significance. This is a very positive move, as should be clear by now.

Beta and Power

Beta is essentially the likelihood that we will fail to detect an effect that happens to be there, that is, the likelihood that we fail to detect a true hypothesis. As we saw, a generally accepted standard for beta is 80 percent (which corresponds to alpha = 0.05). Power is closely related beta. Technically, it is defined as 1- ; so if beta = 80 percent, power = 20 percent.

An important issue to keep in mind is that each statistical test has its own power because each test has a different effect size. That is why there are programs like G*power, which will calculate the power or the **sample size** required to achieve a given power, for each test. So, it does not make sense to say that a given experiment has such and such a power, without also specifying what statistical test you are using.

Effect Size

The effect size is the degree to which your hypothesis is supported. The null hypothesis is the absence of support; in other words, if the null hypothesis turns out to be the best descriptor of the data, then your hypothesis receives no support. In fact it has been said that effect size is "an index of degree of departure from the null hypothesis" (Cohen, 1988, p. 10). Consider again the Winke (2013) study on enhanced versus unenhanced input (words) in reading, which we saw in Chapter 8. Among her results (p. 336; Table 11.1), we saw that enhanced words had a mean total fixation time of 1,475.84 milliseconds, whereas unenhanced words had a mean total fixation time of 1,113.85 milliseconds. Hence the effect size of the manipulation is 361.99 milliseconds. This is called an absolute effect size (it is reported the unit of measurement used in the study, milliseconds in our case). However, a more commonly reported form of effect size is a standardized effect size. What that means is that the effect size is reported in a measurement-independent way. This is very useful, especially when reporting a meta-analysis which aggregates different studies, which may have used different measurement scales. There are various ways of calculating a standardized effect size. One of the simpler is Cohen's *d*, which consists of calculating the difference between the means (as we did above) and then dividing it by the standard deviation (Testing for normality section). Since the pooled standard deviation, that is, the standard deviation of both groups taken together, is 577.69 (Winke, 2013, p. 336; Table 11.1), our standardized effect size is 0.62,[10] which is considered a medium effect.[11] For more examples and other methods to calculate standardized effect sizes, see Sullivan and Feinn (2012) and Cumming (2012) for a thorough treatment.

When you are planning your own study, you will need an effect size prediction to calculate the power of your experiment and thus the number of subjects you need. The place to find out the expected effect size is in the literature review (see Chapter 10). Essentially, one of the reasons for doing a literature review is to see the range of effect sizes that scholars who are looking at data and hypotheses similar to those you are concerned with. If most studies are finding a small effect size, you will need more subjects, to make sure your data are sensitive enough to detect it.

Finally, effect size can answer a crucial question: Will the data you collect answer the question you posed? Consider the following example: suppose you have developed a teaching procedure which improves students' performance on the TOEFL test. It takes 10 hours of classroom instruction, and it has been quite successful in your experience. So, you want to test its effectiveness to know exactly how effective it is. You perform a well-planned randomized study using a large sample of students, with a control group that receives alternative instruction of the same length. You then compare the performance of the two groups and you find, to your delight, that the students in the group instructed with your method do better, and there is a significant difference ($p > 0.05$) between the two. However, before you uncork the champagne bottles you check the effect size and find that it is 0.01. In other words, the instruction results in an increase of the TEOFL score of 1 percent. So, while you did find a significant correlation between the increase in performance and the instruction, the effect is so small that it would probably not be worth implementing it (since the overall score of the TOEFL ranges from 0 to 120, a 1 percent increase would be slightly above 1 point overall). It is in this sense that we mean whether the data answer the question. If your question is "Can I improve the teaching of students wishing to take the TOEFL?" the answer is yes, but it's not worth doing.

Using R

Finally, we briefly discuss R, a software package that has become increasingly popular among applied linguists. What should you use for your statistical analysis? There are hundreds of options, both free or proprietary software packages, such as SPSS (IBM) or Statcrunch (Pearson). The Wikipedia web page comparing the various options is quite helpful: `https://en.wikipedia.org/wiki/Comparison_ of_statistical_packages`. In our classes, we recommend using R.

The first and most obvious reason is that R is free to download and to use. The proprietary statistical packages charge a fee, often a subscription fee, so, in fact you are just renting the software and at the end of the period, your subscription goes away and you are unable to use your hard-won expertise and possibly any files saved to a proprietary format. The second reason is that R is in many ways superior to the proprietary packages. R has a vast and completely free support community on the internet. There are hundreds of tutorials and introductions, ranging from the absolute beginner to extremely advanced subjects. One can literally teach themselves about how to use R entirely using web resources. There are even newsgroups and websites where experts will answer questions from "newbies" and help them figure things out. You can find these resources with a simple search on the internet. Finally, R is disarmingly easy to use, once you figure out how it works. R does not use a graphical user interface (GUI), so there are no pull-down menus or buttons. Instead R relies on the good, old-fashioned command line interface, where the user types a command and the software does the work. What follows is an example, obviously a simple one, but we think it makes nicely our point: the command line, with classical simplicity, tells R to use the hist() function with the variable exam1 (which is where we stored our data) as its argument.

```
hist(exam1)
```

The output is a nice histogram of the data.

Setting R Up

Setting up R is relatively simple. For example, we would recommend downloading the open-source Anaconda package (free, available for all platforms) which includes R (`https://www.anaconda.com`). Another excellent option is R-Studio (`https://rstudio.com`). Alternatively, R can be downloaded as a stand-alone program from the R project web page (`https://www.r-project.org`). There are dozens of videos available online with step-by-step instructions on how to download and install R.

Downloading Packages

While R is an extremely useful program its real power comes from the hundreds of dedicated packages that tackle the specific tests or techniques. A quick search on the internet usually finds a package (and sometimes several) to accomplish a given task. For example, a search for

```
linear mixed models in R
```

shows that the package "lme4" is the go-to package. Generally speaking, even if you are new to this sort of statistical analysis, it is relatively easy to figure out how to apply these packages, because there are many tutorials and videos with worked-out examples. There are even packages for the processing of saccades and fixations from raw data, if you want to do it yourself (most eye trackers will do that for you).

Conclusion

This concludes our introduction to statistical analysis. Let us repeat what we said at the outset: it is impossible to teach you everything there is to know about statistics in a chapter. Our goal here was to provide students without a background in statistics and who might be put off by the traditional mathematical presentation of statistics with enough knowledge to make an educated choice as to which statistical test to choose. Once you determine that, for example, a linear model is the way to go for your study, then you need to become an expert (or at least a passable amateur) on that particular test. That part of the road you need to do by yourself or with the help of a statistician.

Self-Study Questions

The following questions pertain to the contents of this chapter. You should be able to answer them without looking up the answer in the book.

1. Using the chart in Figure 11.7, determine what type of statistical test you would need to use in the following scenarios:
   - **(a)** You measured the number of first, second, and third fixations to three AOIs: eyes, mouth, and rest of the face.
   - **(b)** You measured dwell time on an AOI and you want to correlate it to IQ of the viewers.
   - **(c)** You measured dwell time on an AOI and you want to correlate it to expertise of the viewers (expert, intermediate, beginner).
   - **(d)** You measured whether the speakers look at the AOI (face) when they are being ironical or not.
2. A teacher in an online course notices that the students who attended the live sessions outperformed those who did not. He tests the correlation between attending the lectures and grades and finds that on average, students who attended outperform those who did not attend by almost a full letter grade (so about 10 percent). This is a large effect size, so our teacher is very pleased to find that his lectures are so good that they improve the students' performance so much. Sadly, the test run by the teacher proves no such thing. What confounding variable has he forgotten to consider?

Words to Know

alpha	beta	power	effect size
sample size	type 1 error	type 2 error	parametric test
normal distribution	bell curve	variance	standard deviation
histogram	bar chart	scalar	variable
continuous	binary data	ordinal data	confounding variable
independent variable	dependent variable	blocking variable	explanatory variable
sampling			

Further Reading and Sources

Hatcher (2013) and Field et al. (2012) are broad surveys of statistics. Kachigan (1982) is somewhat out of date but very clear for the fundamentals. Baayen (2008), Johnson (2008), Gries (2013), and Winter (2020) are targeted at linguistics specifically. Fields et al. (2012) and Winter (2020) use R. Winter (2020) provides an introduction to R as well. His presentation of linear models is outstanding.

Conclusions

We can now wrap up our discussion of eye tracking in applied linguistics with a few observations. The most obvious one is that eye tracking as a methodology has produced a body of research of unquestionable significance. It is not an exaggeration to maintain that the impact of eye tracking on psychology, psycholinguistics, and applied linguistics has been transformative (Magnuson, 2019). The evolution of the technology, both on the mechanical side (miniaturization of the mechanical components of the eye-tracking apparatus) and on the analytical side (automatic data handling, fixation and saccade identification, AOI tracking, etc.), has had and will likely continue to have a massive effect on the field. We believe that this will happen in two directions: (1) in the spread of the tool, which is increasingly becoming affordable, so that any department may afford the instrumentation and the day that an individual researcher will own their own eye tracker may not be that remote (in the 1950s few people[1] imagined the spread of personal computers, let alone smartphones); but also (2) in the kind of analyses that are feasible and on the kind of topics that get researched (e.g., the recent emergence of social eye tracking, made practical by the miniaturization of wearable eye trackers). For example, we see as a realistic possibility the deployment of an array of thirty or forty eye-tracking devices in a classroom recording the gaze of all the participants in the interaction. What insight this granularity of data may give us on the dynamics of group interaction is yet to be seen. But, and this is our point, as of 2021, the only thing between this idea and reality is the cost of such a tool and the availability of massive parallel processing/storage devices.

The methodology (or rather the fields employing the methodology) is reaching a certain degree of theoretical maturity, witness the beginning of discussions on the theoretical underpinning or assumptions of the scholarly practices researchers engage in. For example, there is a serious blind spot (pun intended) in the use of eye tracking in applied linguistics: the reliance (implicit or explicit) with the eye-mind theory, which is too vague to be functional as an applied theory of attention, especially in view of much more sophisticated approaches to attention theory, such as perceptual load theory (just to name one) or the work on the neuropsychology of attention

(Cohen, 2014). Two parallel trends also point in the same direction: the increasingly large number of meta-analyses and of replication studies. Making a purely subjective assessment, there seems to be more replication work done in studies involving eye tracking. So perhaps the field of applied linguistics will avoid the replication crisis witnessed in the social sciences (e.g., Bohannon, 2015, for a readable summary).

Finally, we hope to have shown that eye tracking has been successfully applied on a broader scope than reading and language processing: we are witnessing a more inclusive set of phenomena being considered with social eye tracking, which by including (multiparty) conversations, facial expressions, and other "contextual" factors is thus moving toward a more ecological stance. At the same time, the fundamental multimodality of visual input is beginning to be recognized and to figure in the research agenda of scholars. This is especially true of studies in Lx acquisition. Finally, we'd like to point out at the collaborations between linguists and scholars in other fields (such as human-computer interaction) as one of the promising areas where we expect to see increasing research.

For those readers wishing to continue their path of learning in eye tracking (in linguistics and not), we recommend the following books. Bergstrom and Schall (2014) is an approachable introduction to eye tracking, with a broad nonlinguistic focus. Conklin et al. (2018) is more focused on applied linguistics and is geared toward a hands-on approach. Godfroid (2020) is more focused on second language acquisition and bilingualism. Duchowski (2017) and Holmqvist and Andersson (2017) are more advanced and not focused on linguistics.

NOTES

Introduction

1 Aka, Purkinje image or reflection; see Section 1.1. Normally the structure of the eye produces four glints.

Chapter 1

1 The Wikipedia page on the Troxler fading is exceptionally good and contains several demonstrations.

2 Wells does not use the term "saccade," which came into use almost a century later (Wade et al. 2003; Wade 2010, p. 40; 2015).

3 Hering and Hueter wrote in German. Hering uses the word *Rucke*, which translates to "jerk."

4 Very little is known about Lamare, beyond the fact that he was a doctor, worked with Javal, and that he delivered a communication in 1892, in the presence of Javal and other luminaries, which was then printed in the Proceedings of the French society of Ophthalmology. In fact, from the books and documents that mention his work, we don't even know what his first name was: he is usually listed as M. Lamare, but the M. is the abbreviation of *Monsieur* (Mister). We have been unable to locate any further biographical information on him. Both Javal and Tscherning, the successor of Javal at the ophthalmology laboratory, who also quotes Lamare's work, refer to him as "Lamare."

5 Also in 1878, Alex Crum Brown in an article published in *Nature* summarizing work by French doctor Elie de Cyon noted that "the eyes move by jerks" (p. 658).

6 A good account of Stratton's work can be found in Wade (2015).

7 Cocaine was synthesized in 1855 and was used as an anesthetic. In 1884, its use for the anesthesia of the eyes was demonstrated in a famous experiment in which the researcher pricked his own eye with pins, after anesthetizing it with cocaine. Thus, there was nothing unusual in Delabarre's use of it.

8 Chinese white is the name of the pigment, based on zinc oxide.

9 Purkinje images are named after Czech anatomy scholar Jan Evangelista Purkynje (1787–1869), who discovered them. The first Purkinje image is also called the "glint" (as we saw in the introduction).

10 The electric potential can be conceived as the amount of force one would have to exert to move a charge in an electric field. A magnetic field is a good example: if you want to pull apart two magnets you need to exert force on them.

11 The integrated circuit, on which computers are based, was invented in 1958. Computers did not become widespread until the 1970s.

Chapter 2

1 Nowadays, attachments for smartphones exist that can take fundus pictures (and cost less than $100).

Chapter 3

1 In the online version of the book, purple and green, respectively.

2 A nanometer is a billionth of a meter, that is, a 1 preceded by nine zeroes (0.0000000001 meters). To give an idea a sheet of paper is 100,000 nanometers thick.

3 "Ionization" is a process whereby an atom acquires a positive or negative electrical charge by gaining or losing an electron.

4 M cells are also called "parasol cells" due to their shape; the much smaller P cells are called "midget" cells.

5 The Brodmann areas, of which there are over forty in humans, were identified and labeled by a German anatomy researcher by that name in 1909, on the basis of the nature of the cells. Later studies have identified as many as 107 areas. Brodmann areas remain a frequently used system of referencing locations in the brain.

6 Hubel is obviously joking: from an evolutionary standpoint, it is perfectly logical that attention to movement would evolve long before attention to static objects does: a predator is likely to move, as it tries to eat you, but a tree trunk does neither of these things.

7 Technically, the two halves of the brain are *not* entirely symmetrical; however, the differences are relatively small and not well understood, in terms of effect on observable behavior (Tzourio-Mazoyer & Seghier, 2016).

8 ERP stands for event-related potential and is a record of the electrical activity of the brain. N400 means that 400 milliseconds from the onset of the stimulus there is a negative (N) wave. N400 waves are associated with incongruity.

Chapter 4

1 The math is simple: we have eight samples 0, 1, 2, . . ., 7; each sample is taken at 3.3 ms intervals (i.e., a sampling rate of 300 Hz), hence 8 × 3.3 = 26.4 ms.

2 There are other, more complex, approaches based on mathematical techniques and statistical approaches beyond the threshold and dispersion models discussed in the text. Those are outside of the purview of an introductory text. See Salvucci and Goldberg (2000) for some discussion.

3 Nonpsychological applications of eye tracking exist: eye tracking may be used for medical/physiological studies and for purely engineering applications, as in HCI.

4 Sincere smiles are defined as smiles involving the rising of the corners of the mouth and the wrinkling of the corner of the eye.

5 However, you should refrain from doing so *after* you have started analyzing the data.

6 The translated caption reads: "When you use legal loopholes in your favor."

7 This is true, generally speaking, but it is not always so. If pupillometry is part of your goals and you are selecting an eye tracker, it is wise to check if the pupil size data are easily retrievable.

Chapter 5

1 For example, Zang et al. (2011) reported a shorter perceptual span in Chinese; Paterson et al. (2014) reported that in Urdu, which reads right to left, the asymmetry of the perceptual span is present but in the opposite direction as English.

2 We will not address functional illiteracy.

3 The reader will recall that phonemes are written between slashes: /a/ = the phoneme /a/; [a] = the allophone [a].

4 A "fixed-width" font is a font, like Courier, in which all letters have the same width. They were developed for computer screens. Typewriters are also fixed-width.

5 The correlation between length and frequency is known as Zipf's law.

6 In fact, it's not happenstance; the principle of least effort ensures that the most frequent words get shortened for ease of production.

7 Actually, the most frequent words of English are "the," "a," "is," "I," "in," and so on, but these tend to be function words and thus are generally skipped during reading.

8 Length is tied to frequency by the principle of least effort (aka, Zipf's law). One of its consequences is that the most frequent words are also the shortest (as the speakers seek to minimize effort).

9 The cloze test was introduced by Taylor (1953) and comes from Gestalt psychology where "closure" indicates the perception of separate parts as a whole so that, for example, if a circle has several missing parts, we nonetheless perceive it as a circle.

10 The original studies that postulated this explanation were done using neural networks, so this is a broadening of the sense in which they originally used the term, meaning plasticity of the neural network.

11 We know this from various psycholinguistic studies, for example, Onifer and Swinney (1981).

12 The reason that "eye" is considered irregular is that it violates the sound-to-grapheme rules of English. There are no, or very few, other words that sound [ai] and are spelled "eye." Conversely, there are plenty of words in English that use the "oo" spelling for the sound [u] (hence it is regular).

Chapter 6

1 Cf. Marslen-Wilson (1987); this model is known as the cohort activation model (see later in the text).

2 This model may strike you as oddly unrealistic, but it was in fact the central tenet of generative-transformational grammar, especially as practiced by Chomsky and his followers. There were, needless to say, many other scholars who objected to some or all of this idea.

3 The original term was "word-initial cohort" (Marslen-Wilson & Tyler, 1980), which is clearer, but the term "cohort" has become standard.

4 A confederate is a member of the experimental team who pretends to be a regular participant in the experiment. The use of confederates presupposes a deception protocol and is subject to special regulations.

5 There is a small class of postmodifying adjectives, such as "elect" and "designate," but they can be ignored in this context.

6 Contrastive stress is a prosodic marker indicating that the stressed item is contrasted against another item from a class inferred from context. For example, if a speaker utters "No, I don't want CHOMSKY's book," they do want a book, just not one by Chomsky. Contrastive stress is marked in English by high volume and high pitch.

7 If we remove the contrastive stress, the disambiguation point is *red*, since there are two figures for which the color disambiguates: the large red square and the large blue circle. The presence of the contrastive stress forces the participants to focus on the red pair, since there is no alternative small blue circle or large yellow triangle.

8 German has a different intonation for SVO sentences and OVS sentences. Both are grammatical.

9 Disfluencies are hesitations, false starts, filled pauses, word repetitions, slips of the tongue, and other errors, during speech.

10 The Stroop effect is named after J. R. Stroop who published a widely quoted 1935 article. However, the effect was already documented in prior literature (MacLeod, 1991, p. 163) going as far back as Wundt.

11 The original experiments that Stroop performed (1935) are more extended: Stroop examines incongruent words and words written in black. See MacLeod (1991).

12 A nice video exemplifying the McGurk effect can be found here: `https://www.youtube. com/watch?v=2k8fHR9jKVM`.

Chapter 7

1 The term "triadic" for an interaction between three people is to be avoided, due to an ambiguity with the use of "triadic" in HCI, which predates it. We will use the term "three-person interaction" in such cases.

2 Lexical items clearly have socially codified meanings; syntactic and semantic rules that assemble sentences also have more or less stable meanings. Implicatures are more open ended, but nonetheless an inferential path *can* be identified so that the intention of the speaker to imply something can be reconstructed (Grice, 1989). On implicatures, see also Chapter 6. Conversely, contextualization cues are mostly completely subconscious and there is no grammatically codified meaning attached to them (Gumperz, 2015).

3 ELAN (2020) is an open-source software platform for annotating and transcribing audio and video files developed at The Language Archive (Max Planck Institute for Psycholinguistics), over the last twenty years. This requires a marker (e.g., a sound) that occurs in the two data streams and that can be used as a fixed point to align the files. ELAN is widely used for video annotation/transcription, but there are other options such as EXMARaLDA (`https: //exmaralda.org/en/`) or ATLAS.ti (`https://atlasti.com`), and more.

4 Possibly one could have two virtual agents talking to each other.

5 Apparently, blinking counts as "breaking" the gaze in a staring contest. For a study of cooperative and competitive staring contests, see Jarick and Kingstone (2015).

6 See also Carletta et al. (2010) and Clark and Gergle (2011) who also used the term "dual eye tracking" independently of Jermann et al.'s use. The earliest dual eye-tracking setup we have been able to locate, in the context of HCI, was Bailly et al. (2006), who did not use the term "dual eye tracking," however. The earliest uses of "social eye tracking" we have tracked down were Schneider et al. (2013) and Cheng et al. (2015). The earliest social eye-tracking setup we have identified, which used three speakers at a time, is Kabashima et al. (2012). Kabashima et al. did not use either the terms dual or social eye tracking but referred to their data as "multiparty" and "multimodal."

7 Tetris was a very popular video game in which falling blocks had to be arranged to minimize the number of gaps between the pieces. The original game was single-player, but Jermann et al. modified it to make it a two-player game.

8 A good eye tracker can certainly make the distinction when the speakers are at relatively close distance. However, people do not seem to be very good at making such distinctions; as Kleinke stated, "it appears that outside of very special viewing conditions, the discrimination between eye-gaze and face-gaze

is not easily made" (p. 79). If the speakers are very close (less than 1 meter, or about 3 feet) accuracy obviously improves (Argyle & Cook, 1976, p. 47).

9 https://www.robopec.com/en/constructions/others/reeti-robopec/.

10 A video of the robot being ironical can be found at the following link: https://www.youtube.com/watch?v=ZeN37TDuUCE&feature=youtu.be.

11 The surprising nature of this finding is of course dispelled by a simple consideration: gaps between turns are in the order of 200 ms. It takes 500 ms for a stimulus to reach consciousness, and language production takes over 600 ms and as long as 1,500 ms (Levinson and Torreira, 2015). Hence, the only possible explanation is that the speakers *have already decided what they want to say* before the previous turn is over.

12 "Transition-relevant point" is a technical term from conversation analysis, meaning a point in the conversation where speakers feel that a change in speaker may take place.

Chapter 8

1 The term "second language" acquisition refers to the learning of a language other than the native one. The terminology has shifted in recent years, in recognition of the fact that most speakers are bilingual or multilingual and may have grown up speaking several languages. It is in fact monolingualism, as, for example, is the experience of many native speakers of English in the United States, that is exceptional. We now prefer L1 and Lx, since the language acquired non-natively may be a third, fourth, or fifth one. We will continue to use the term SLA to describe the field as a whole because of its widespread acceptance.

2 The difference is between stopping your task to think "I am hungry" and taking your attention partially off the task.

3 Considering that the *longest* video is less than 11 minutes long, this is quite a remarkable claim.

4 Presumably the students made up the lost information by studying. As Sherlock Holmes famously said, once you have eliminated all alternatives, the remaining hypothesis must be true, however improbable it may seem.

5 The novel *Sense and Sensibility* was read in its entirety, in 12- to 15-hour-long sessions, over multiple days.

6 Our emphasis, SA/LP.

7 We ignore, for simplicity, coarticulation effects. For example, the presence of an [i] vowel *after* a consonant could be predicted by the palatalized coarticulation of the consonant.

8 Cooper et al. (2002) already considered Dutch Lx speakers of English as well.

9 As in "John made his dog a little house."

10 RAGE stands for Reduced Ability to Generate Expectations (Grüter et al., 2014).

11 It is disappointing that Kaan herself did not coin the RAGE term, as it would have been known as the Rage of Kaan.

12 The sentence is a garden path (see Chapter 5). When a reader processes it, it starts out as "Put the frog on the napkin," where the napkin is the location to which the frog has to be moved, but the occurrence of second prepositional phrase "into the box" forces a reinterpretation of "on the napkin" as the location of the frog and not its destination.

13 The difference between the two is that captions are in the same language as the spoken version of the text; subtitles are in a different language, or more than one.

14 It should be noted that broadcasting with closed captioning began in 1979 in the UK (BBC) and in 1980 in the United States. It was not until 1991 that TV sets were mandated to have closed captioning capacity.

Chapter 9

1 For example, `https://www.youtube.com/watch?v=ID5GjiLhDDA`.

2 The series is characterized by large spreads depicting very crowded scenes with people doing various things in a location. In the scene there is always one character (Waldo) and part of the goal is to find Waldo.

3 A similar map is also found in the frontal eye field, an area of the frontal cortex that determines and controls top-down saccades, that is, saccades directed by goals, or semantic information.

4 Friedman (1979) presented the idea of the gist of an image within frame theory, in the context of recall. Experimental subjects recall the "themes of pictures" (i.e., their gist) rather than the actual visual details, hence explaining the difficulty in identifying when an image has been modified, unless the change is incongruous.

5 It should be noted that Liechty et al. (2003) did not use a fixation cross and the participants freely leafed through in a magazine-like booklet. Hence the choice to start at the center is not affected by the experimental setup. The setup used by Liechty et al. (2003) is described in detail in Wedel and Pieters (2000).

6 For example, position of the nose, hair position, and color, and so on.

7 The point of expansion is the point in the visual field of a subject in motion in which objects appear not to be moving; this is not the vanishing point of perspective, although it is a similar construct.

8 Whereas in the point of expansion/tangent point, the moving subject focuses on a point ahead such that the distance between the point and the subject does not change, in the waypoint hypothesis, the moving subject focuses on a specific point in the road ahead, which then approaches and eventually is passed by the subject. Summing up, the point of expansion does not move, relative to the moving subject, whereas the waypoint does move, relative to the subject.

9 A possible explanation of this fact is that the ads in Pieters et al. (2007) appeared in the context of a newspaper page, and hence with other ads, whereas the ads in the Pieters and Wedel (2004) were single-page magazine ads and there the text may have been more important, since the reader already is paying attention to the page and their attention need not be secured (Higgins et al., 2014).

10 No mention is made in Nielsen and Pernice (2010) of cats. We were not able to find data about preferences for images of cats or horses. We did however come across studies of canine and feline eye tracking. Those are, however, outside the scope of this book.

11 The technical definition is "the extent to which a system, product or service can be used by specified users to achieve specific goals with effectiveness, efficiency and satisfaction in a specified context of use" (ISO 9241:11:2018, International Organization for Standardization). `https:// www.iso.org/obp/ui/#iso:std:iso:9241:-11:ed-2:v1:en`.

12 It is significant that the only source out of several hundred we consulted for the writing of this book that discusses in any detail the cost of an eye-tracking study is Nielsen and Pernice (2010). This is all the more remarkable since the affordability of eye-tracking equipment has clearly influenced its growth as a field. To give an idea of how much the costs have dropped, for the price of what Nielsen and Pernice paid to rent the eye-tracking equipment for five months, they could easily buy one outright today.

Chapter 10

1 On qualitative methods in applied linguistics, see Heigham and Croker (2009).

2 In those countries where Google Scholar is not available, there may exist alternatives, such as Baidu scholar in China (`https://xueshu.baidu.com`).

3 In our experience, editors of volumes are systematically left out of references for chapters in edited books; often page numbers are missing; sometimes the reference is flat out wrong. In short, you should always double check a reference found online.

4 You may wonder about originality and innovation: won't the weight of prior research approaches stifle the originality and creativity of your work? It might, but a truly original thinker will always emerge, regardless of the weight of tradition. If nothing else, it's useful to know what one is pushing against.

5 Always be careful about making that sort of claim. A study that you ignored, published in an obscure journal, usually emerges, days after publication of your paper, to significant embarrassment.

6 This is very important for the scientific standing of your work. You should strive to make falsifiable claims. On falsifiability, see Popper (1959). Falsificationism is now considered somewhat naive but remains the foundation of the scientific method.

7 We will not consider unaffiliated scholars, as the likelihood of an unaffiliated scholar being able to afford an eye tracker, despite the drops in price of the equipment, is still very low. Regardless of affiliation, one does not want to perform unethical research, and one should follow the recommendations of one's country. Moreover, most journals require a statement concerning the ethical treatment of participants and you may be unable to publish your research if you cannot document that you followed a protocol that ensured the safety and well-being of your participants.

8 In the UK the Central Office for Research Ethics Committees (COREC; `http://www.corec. org.uk/`) oversees institutional research ethics committees (REC). In Australia, the National Health and Medical Research Council oversees human research ethics committees (HREC; `http: //www.health.gov.au/nhmrc/issues/index.htm`). In New Zealand the Health Research Council oversees Institutional Ethics Committees (IEC; `http://www.hrc.govt.nz/ethicgui.` htm). We have limited ourselves to Anglophone countries, but of course all countries are likely to have such an institution. You should be able to locate the relevant entity by searching your university website or asking experienced researchers at your institution.

9 Usually, desks or filing cabinets have a lock. If none are available, you can purchase a small lock box for under 20 dollars.

10 Mr. Musk is the second richest person on earth, at the time of writing. Hence he is significantly richer than most people.

11 The size of a stimulus's image on the retina is determined both by its actual size and its distance from the eye.

12 If you are a humanist, you may find this distinction faintly ridiculous, as obviously the way the questions have been formulated, the data gathered, the subjects selected, and so on will "taint" the supposed objectivity of the data. After all, there are no data in nature; data exist only if someone has decided that something counts as a datum. This is all true but irrelevant to the scientific enterprise. The goal of science is not to seek absolute truth but to produce replicable, factual information, following a given methodology.

Chapter 11

1 Not to be macabre, but if you started on Monday at 8.00 am, almost 3,900 of your subjects would have died by 8.00 pm.

2 The terminology is confusing, to say the least: Wikipedia lists a host of other terms; for the independent variable: explanatory variable, predictor variable, regressor, covariate, controlled variable, manipulated variable, input variable, and so on. For the dependent variable: response variable, regressand, criterion, predicted variable, measured variable, explained variable, experimental variable, responding variable, outcome variable, output variable, or label.

3 Causality is another can of worms, but we will leave that one alone for the time being.

4 In this case, the causality is very clear: it is impossible that differences in reading times cause one to be a native speaker, so the causality flows only in one direction.

5 The size of the bins can be changed as needed.

6 Technically, the formula given in the text is the variance of the population; for the variance of the sample, you need to subtract 1 from the number of observations, but we can ignore this in this context. In particular for large samples this makes very little difference. For more detail, see Kachigan (1982, pp. 57–8).

7 Your name is Bond, James Bond.

8 `https://aviation-safety.net/graphics/infographics/ Fatal-Accidents-Per-Mln-Flights-1977-2017.jpg`.

9 The term "p-value" comes from the letter "P," which was used to denote the chi-squared distribution. P-value was first introduced by Karl Pearson, in 1900, using precisely the chi-squared distribution.

10 Just to be completely clear, the math is as follows:

$$\frac{1475.84-1113.85}{577.69} = 0.62$$

11 The scale goes as follows:

Small	0.2
Medium	0.5
Large	0.8
Very large	1.3

Conclusions

1 The famous quote by Thomas J. Watson, the founder of IBM, that the computer market in the United States was limited to five units is apocryphal. However, other people expressed identical ideas: for example, Howard Aiken, one of the pioneers of computing, stated that the United States needed "half a dozen large computers" (Cohen, 1999, p. 292).

BIBLIOGRAPHY

Abdel Latif, M. M. (2019). Eye-tracking in recent L2 learner process research: A review of areas, issues, and methodological approaches. *System*, *83*, 25–35.

Abdessalem, H. B., Chaouachi, M., Boukadida, M., & Frasson, C. (2019). Toward real-time system adaptation using excitement detection from eye tracking. In International Conference on Intelligent Tutoring Systems (pp. 214–223). Cham: Springer.

Abratt, R., & Goodey, S. D. (1990). Unplanned buying and in-store stimuli in supermarkets. *Managerial and Decision Economics*, *11*(2), 111–121.

Alemdag, E., & Cagiltay, K. (2018). A systematic review of eye tracking research on multimedia learning. *Computers & Education*, *125*, 413–428.

Allopenna, P. D., Magnuson, J. S., & Tanenhaus, M. K. (1998). Tracking the time course of spoken word recognition using eye movements: Evidence for continuous mapping models. *Journal of Memory and Language*, *38*(4), 419–439.

Altmann, G. T. (2011). The mediation of eye movements by spoken language. In Liversedge, S. P., Gilchrist, I., & Everling, S. (Eds.). *The Oxford handbook of eye movements*, (pp. 979–1003). Oxford University Press.

Altmann, G. T., & Kamide, Y. (1999). Incremental interpretation at verbs: Restricting the domain of subsequent reference. *Cognition*, *73*(3), 247–264.

Al-Wabil, A., ElGibreen, H., George, R. P., & Al-Dosary, B. (2010). Exploring the validity of learning styles as personalization parameters in elearning environments: An eyetracking study. In 2010 2nd International Conference on Computer Technology and Development (pp. 174–178). IEEE.

Amati, F., & Brennan, S. E. (2018). Eye gaze as a cue for recognizing intention and coordinating joint action. In Brône, G. & Oben, B. (Eds.). *Eye-tracking in interaction: Studies on the role of eye gaze in dialogue*, (pp. 21–46). Benjamins.

Anderson, J. R. (2002). Spanning seven orders of magnitude: A challenge for cognitive modeling. *Cognitive Science*, *26*(1), 85–112.

Antes, J. R. (1974). The time course of picture viewing. *Journal of Experimental Psychology*, *103*(1), 62–70.

Anwar, S. A. (2019). Real time facial expression recognition and eye gaze estimation system. Doctoral dissertation, University of Arkansas at Little Rock.

Anwar, S., Milanova, M., Abdulla, S., & Svetleff, Z. (2018). Emotion recognition and eye gaze estimation system: EREGE. In International Conference on Human-Computer Interaction (pp. 364–371). Cham: Springer.

Arai, M., Bard, E. G., & Hill, R. (2009). Referring and gaze alignment: Accessibility is alive and well in situated dialogue. Proceedings of the Annual Meeting of the Cognitive Science Society (Vol. 31, p. 31).

Argyle, M. & Cook, M. (1976). *Gaze and mutual gaze*. Cambridge.

Arnold, J. E., Fagnano, M., & Tanenhaus, M. K. (2003). Disfluencies signal theee, um, new information. *Journal of Psycholinguistic Research*, *32*(1), 25–36.

Arnold, J. E., Tanenhaus, M. K., Altmann, R. J., & Fagnano, M. (2004). The old and thee, uh, new: Disfluency and reference resolution. *Psychological Science*, *15*(9), 578–582.

Aryadoust, V., & Ang, B. H. (2019). Exploring the frontiers of eye tracking research in language studies: A novel co-citation scientometric review. *Computer Assisted Language Learning*, *34*(7), 1–36.

Ashby, J., & Clifton, C. Jr. (2005). The prosodic property of lexical stress affects eye movements during silent reading. *Cognition*, *96*(3), B89–B100.

Atalay, A. S., Bodur, H. O., & Rasolofoarison, D. (2012). Shining in the center: Central gaze cascade effect on product choice. *Journal of Consumer Research*, *39*(4), 848–866.

Attardo, S. (2020). Scripts, frames, and other semantic objects. In Attardo, S. (Ed.). *Script-based semantics: Foundations and applications. Essays in honor of Victor Raskin*, (pp. 11–41). Mouton De Gruyter.

Attardo, S. & Pickering, L. (2021). *Pragmatics and its applications to TESOL and SLA*. Wiley Blackwell.

Auer, P. (2018). Gaze, addressee selection and turn-taking in three party interaction. In Brône, G. & Oben, B. (Eds.). *Eye-tracking in interaction: Studies on the role of eye gaze in dialogue*, (pp. 197–231). Benjamins.

Auer, P. (2021). Turn-allocation and gaze: A multimodal revision of the "current-speaker-selects-next" rule of the turn-taking system of conversation analysis. [in press] *Discourse Studies*.

Baayen, R. H. (2008). *Analyzing linguistic data: A practical introduction to statistics using R*. Cambridge.

Babcock, J. S., Lipps, M., & Pelz, J. B. (2002). How people look at pictures before, during, and after scene capture: Buswell revisited. In *Human vision and electronic imaging VII*, (pp. 34–47). International Society for Optics and Photonics. Vol. 4662.

Bailly, G., Elisei, F., Raidt, S., Casari, A., & Picot, A. (2006). Embodied conversational agents: computing and rendering realistic gaze patterns. In Pacific-Rim Conference on Multimedia (pp. 9–18). Berlin, Heidelberg: Springer.

Bailly, G., Mihoub, A., Wolf, C., & Elisei, F. (2018). Gaze and face-to-face interaction. In Brône, G. & Oben, B. (Eds.). *Eye-tracking in interaction: Studies on the role of eye gaze in dialogue*, (pp. 139–168). Benjamins.

Baimel, A., Severson, R. L., Baron, A. S., & Birch, S. A. (2015). Enhancing "theory of mind" through behavioral synchrony. *Frontiers in Psychology*, *6*(870), 1–6.

Ballard, D. H., Hayhoe, M. M., & Pelz, J. B. (1995). Memory representations in natural tasks. *Journal of Cognitive Neuroscience*, *7*(1), 66–80.

Balota, D. A., Pollatsek, A., & Rayner, K. (1985). The interaction of contextual constraints and parafoveal visual information in reading. *Cognitive Psychology*, *17*(3), 364–390.

Balota, D. A., Yap, M. J., & Cortese, M. J. (2006). Visual word recognition: The journey from features to meaning (a travel update). In Traxler, M. J. & Gernsbacher, M. A. (Eds.). *Handbook of psycholinguistics*, (pp. 285–375). Academic Press.

Baltrusaitis, T., Zadeh, A., Lim, Y. C., & Morency, L. P. (2018). Openface 2.0: Facial behavior analysis toolkit. In 2018 13th IEEE International Conference on Automatic Face & Gesture Recognition (FG 2018) (pp. 59–66). IEEE.

Barr, D. J., & Keysar, B. (2006). Perspective-taking and the coordination of meaning in language use. In Traxler, M. J. & Gernsbacher, M. A. (Eds.). *Handbook of psycholinguistics* (pp. 901–938). Amsterdam: Elsevier.

Barreto, A. M. (2013). Do users look at banner ads on Facebook?. *Journal of Research in Interactive Marketing*, 7(2), 119–139.

Barthel, M., & Levinson, S. C. (2020). Next speakers plan word forms in overlap with the incoming turn: evidence from gaze-contingent switch task performance. *Language, Cognition and Neuroscience*. 1–20. DOI: 10.1080/23273798.2020.1716030

Bates, R., Istance, H., Oosthuizen, L., and Majaranta, P. (2005). D2.1 survey of De-Facto standards in eye tracking. communication by gaze interaction (COGAIN), IST-2003-511598: Deliverable 2.1. http://wiki.cogain.org/images/1/1b/COGAIN-D2.1.pdf

Bavelas, J. B. (2007). Face-to-face dialogue as a micro-social context. The example of motor mimicry. In Duncan, S., Levy, E., & Cassell, J. (Eds.). *Gesture and the dynamic dimension of language*, (pp. 127–146). Amsterdam: John Benjamins.

Bavelas, J. B., Black, A., Lemery, C. R., & Mullett, J. (1986). "I show how you feel": Motor mimicry as a communicative act. *Journal of Personality and Social Psychology*, *50*(2), 322.

Bavelas, J. B., & Chovil, N. (2000). Visible acts of meaning: An integrated message model of language in face-to-face dialogue. *Journal of Language and Social Psychology*, *19*(2), 163–194.

Bavelas, J. B., Coates, L., & Johnson, T. (2002). Listener responses as a collaborative process: The role of gaze. *Journal of Communication*, *52*(3), 566–580.

Bavelas, J., Gerwing, J., Sutton, C., & Prevost, D. (2008). Gesturing on the telephone: Independent effects of dialogue and visibility. *Journal of Memory and Language*, *58*(2), 495–520.

Bayliss, A. P., Paul, M. A., Cannon, P. R., & Tipper, S. P. (2006). Gaze cuing and affective judgments of objects: I like what you look at. *Psychonomic Bulletin & Review*, *13*(6), 1061–1066.

Becker, W., & Jürgens, R. (1979). An analysis of the saccadic system by means of double-step stimuli. *Vision Research*, *19*, 967–983.

Bednarik, R. (2012). Expertise-dependent visual attention strategies develop over time during debugging with multiple code representations. *International Journal of Human-Computer Studies*, *70*(2), 143–155.

Bergstrom, J. R., & Schall, A. (Eds.). (2014). *Eye tracking in user experience design*. Elsevier.

Behrmann, M., Avidan, G., Thomas, C., & Nishimura, M. (2011). Impairments in face perception. In Rhodes, G., & Haxby, J. (Eds.). *Oxford handbook of face perception*. Oxford University Press.

Beymer, D., Russell, D., & Orton, P. (2008). An eye tracking study of how font size and type influence online reading. People and computers XXII: culture, creativity, interaction. Proceedings of HCI 2008 (pp. 15–18). British Computer Society .

Bialystok, E., Craik, F. I. M., Klein, R., & Viswanathan, M. (2004). Bilingualism, aging, and cognitive control: Evidence from the Simon task. *Psychology and Aging*, *19*(2), 290–303.

Bicknell, K., & Levy, R. (2010). A rational model of eye movement control in reading. In Proceedings of the 48th Annual Meeting of the Association for Computational Linguistics (pp. 1168–1178).

Biga, L. M., Dawson, S., Harwell, A., Hopkins, R., Kaufmann, J., LeMaster, M., Matern, P., Morrison-Graham, K., Quick, D. Runyeon, J., & Whittier, L. (2019). *Anatomy & physiology*. Corvallis: Oregon State University.

Bixler, R., & D'Mello, S. (2016). Automatic gaze-based user-independent detection of mind wandering during computerized reading. *User Modeling and User-Adapted Interaction*, *26*(1), 33–68.

Blais, C., Jack, R. E., Scheepers, C., Fiset, D., & Caldara, R. (2008). Culture shapes how we look at faces. *PloS One*, *3*(8), e3022.

Blignaut, P. (2014). Mapping the pupil-glint vector to gaze coordinates in a simple video-based eye tracker. *Journal of Eye Movement Research*, *7*(1). 1–11.

Bock, K., Irwin, D. E., & Davidson, D. J. J. (2004). Putting first things first. In Henderson, J. M., & Ferreira, F. (Eds.). *The interface of language, vision, and action: Eye movements and the visual world*, (pp. 249–278). Psychology Press.

Bock, K., & Ferreira, V. (2014). Syntactically speaking. In Goldrick, M., Ferreira, V., & Miozzo, M. (Eds.). *Oxford library of psychology. The Oxford handbook of language production* (pp. 21–46). Oxford University Press.

Bohannon, J. (2015). Many psychology papers fail replication test: An effort to repeat 100 studies yields sobering results, but many researchers are positive about the process. *Science*, *349*(6251), new series, 910–911.

Bolger, C., Bojanic, S., Sheahan, N., Malone, J., Hutchinson, M., & Coakley, D. (2000). Ocular microtremor (OMT): A new neurophysiological approach to multiple sclerosis. *Journal of Neurology, Neurosurgery, and Psychiatry*, *68*(5), 639–642. https://doi.org/10.1136/jnnp.68.5.639

Brancucci, A., Lucci, G., Mazzatenta, A., & Tommasi, L. (2009). Asymmetries of the human social brain in the visual, auditory and chemical modalities. *Philosophical Transactions of the Royal Society B: Biological Sciences*, *364*(1519), 895–914.

Breen, M., & Clifton Jr., C. (2011). Stress matters: Effects of anticipated lexical stress on silent reading. *Journal of Memory and Language*, *64*(2), 153–170.

Brennan, S. E., Hanna, J. E., Zelinsky, G. J., & Savietta, K. J. (2012). Eye gaze cues for coordination in collaborative tasks. DUET 2012 Workshop: Dual eye tracking in CSCE. 2012 ACM Conference on Computer Supported Cooperative Work, Seattle, WA. 1–8.

Brishtel, I., Khan, A. A., Schmidt, T., Dingler, T., Ishimaru, S., & Dengel, A. (2020). Mind wandering in a multimodal reading setting: Behavior analysis & automatic detection using eye-tracking and an EDA sensor. *Sensors*, *20*(9), 2546.

Brône, G., & Oben, B. (Eds.). (2018). *Eye-tracking in interaction: Studies on the role of eye gaze in dialogue*. Benjamins.

Brown-Schmidt, S., & Konopka, A. E. (2008). Little houses and casas pequeñas: Message formulation and syntactic form in unscripted speech with speakers of English and Spanish. *Cognition*, *109*(2), 274–280.

Buswell, G. T. (1935). *How people look at pictures: A study of the psychology of perception in art*. Chicago: The University of Chicago press.

Butcher, K. R. (2014). The multimedia principle. In Mayer, R. E. (Ed.). *The Cambridge handbook of multimedia learning*, (pp. 174–205). Cambridge.

Calder, A., Rhodes, G., Johnson, M., & Haxby, J. (2011). *Oxford handbook of face perception*. Oxford University Press.

Campbell, M. J. & Shantikumar, S. (2016). Statistical methods. https://www. healthknowledge.org.uk/public-health-textbook/research-methods/1b-statistical-methods

Carletta, J., Hill, R. L., Nicol, C., Taylor, T., De Ruiter, J. P., & Bard, E. G. (2010). Eyetracking for two-person tasks with manipulation of a virtual world. *Behavior Research Methods*, *42*(1), 254–265.

Carrol, G., & Conklin, K. (2017). Cross language lexical priming extends to formulaic units: Evidence from eye-tracking suggests that this idea 'has legs'. *Bilingualism: Language and Cognition*, *20*(2), 299–317.

Carrol, G., & Conklin, K. (2020). Is all formulaic language created equal? Un-packing the processing advantage for different types of formulaic sequences. *Language and Speech*, *63*(1), 95–122.

Carroll, P., & Slowiaczek, M. L. (1986). Constraints on semantic priming in reading: A fixation time analysis. *Memory & Cognition*, *14*(6), 509–522.

Carter, B. T., & Luke, S. G. (2020). Best practices in eye tracking research. *International Journal of Psychophysiology*, *155*, 49–62.

Chabris, C. F., & Simons, D. J. (2010). *The invisible gorilla: And other ways our intuitions deceive us*. New York: Random House.

Chambers, C. G., Tanenhaus, M. K., Eberhard, K. M., Filip, H., & Carlson, G. N. (2002). Circumscribing referential domains during real-time language comprehension. *Journal of Memory and Language*, *47*(1), 30–49.

Chambers, C. G., Tanenhaus, M. K., & Magnuson, J. S. (2004). Actions and affordances in syntactic ambiguity resolution. *Journal of* Experimental Psychology: *Learning, Memory, and Cognition*, *30*(3), 687–696.

Chandon, P., Hutchinson, J. W., Bradlow, E. T., & Young, S. H. (2009). Does in-store marketing work? Effects of the number and position of shelf facings on brand attention and evaluation at the point of purchase. *Journal of Marketing*, *73*(6), 1–17.

Chapman, P. R., & Underwood, G. (1998). Visual search of dynamic scenes: Event types and the role of experience in viewing driving situations. In Underwood, G. (Ed.). *Eye guidance in reading and scene perception*, (pp. 369–393). Amsterdam: Elsevier.

Charlier, J., Buquet, C., Dubus, F., Hugeux, J. P., & Degroc, B. (1997). VI-SIOBOARD: a new gaze command system for handicapped subjects. *Medical and Biological Engineering and Computing*, *35*, 461–462.

Cheng, S., Sun, Z., Ma, X., Forlizzi, J. L., Hudson, S. E., & Dey, A. (2015). Social eye tracking: Gaze recall with online crowds. In Proceedings of the 18th ACM Conference on Computer Supported Cooperative Work & Social Computing (pp. 454–463).

Choi, Y. S., Mosley, A. D., & Stark, L. W. (1995). String editing analysis of human visual search. *Optometry and Vision Science*, *72*(7), 439–451.

Christensen, L. B., Johnson, B., & Turner, L. A. (2014). *Research methods, design, and analysis*. Pearson.

Cieślicka, A. B., Heredia, R. R., & Olivares, M. (2014). It's all in the eyes: How language dominance, salience, and context affect eye movements during idiomatic language processing. In Pawlak, M. & Aronin, L. (Eds.). *Essential topics in applied linguistics and multilingualism* (pp. 21–41). Cham: Springer.

Clark, A. T., & Gergle, D. (2011). Mobile dual eye-tracking methods: challenges and opportunities. In Proceeding of International Workshop on Dual Eye Tracking.

Clark, H. H. (1996). *Using language*. Cambridge University Press.

Clark, R. C., & Mayer, R. E. (2016). *E-learning and the science of instruction: Proven guidelines for consumers and designers of multimedia learning*. Wiley.

Clifton Jr, C., & Staub, A. (2008). Parallelism and competition in syntactic ambiguity resolution. *Language and Linguistics Compass*, 2(2), 234–250.

Coderre, E. L., Van Heuven, W. J., & Conklin, K. (2013). The timing and magnitude of Stroop interference and facilitation in monolinguals and bilinguals. *Bilingualism: Language and Cognition*, *16*(2), 420–441.

Cognolato, M., Atzori, M., & Müller, H. (2018). Head-mounted eye gaze tracking devices: An overview of modern devices and recent advances. *Journal of Rehabilitation and Assistive Technologies Engineering, 5*, 2055668318773991.

Cohen, I. B. (1999). *Howard Aiken: Portrait of a computer pioneer*. MIT Press.

Cohen, J. (1988). *Statistical power analysis for the behavioral sciences*. Lawrence Earlbaum.

Cohen, R. A. (2014). *The neuropsychology of attention* (2nd ed.). Springer.

Collins, J. B. (2017). Online dictionaries and the involvement load hypothesis: An empirical study. *APU Journal of Language Research*, 2, 33–49.

Conati, C., Aleven, V., & Mitrovic, A. (2013). Eye-tracking for student modelling in intelligent tutoring systems. In Sottilare, R. A., Graesser, A., Hu, X., & Holden, H. (Eds.). *Design recommendations for intelligent tutoring systems*, (pp. 227–236). U.S. Army Research Laboratory.

Conati, C., & Merten, C. (2007). Eye-tracking for user modeling in exploratory learning environments: An empirical evaluation. *Knowledge-Based Systems*, *20*(6), 557–574.

Conklin, K., Alotaibi, S., Pellicer-Sánchez, A., & Vilkaite-Lozdiene, L. (2020). What eye-tracking tells us about reading-only and reading-while-listening in a first and second language. *Second Language Research, 36*(3). 257–276.

Conklin, K., Pellicer-Sánchez, A., & Carrol, G. (2018). *Eye-tracking: A guide for applied linguistics research*. Cambridge.

Conley, Q., Earnshaw, Y., & McWatters, G. (2020). Examining course layouts in blackboard: Using eye-tracking to evaluate usability in a learning management system. *International Journal of Human–Computer Interaction*, *36*(4), 373–385.

Connell, K., Hüls, S., Martínez-García, M. T., Qin, Z., Shin, S., Yan, H., & Tremblay, A. (2018). English learners' use of segmental and suprasegmental cues to stress in lexical access: An eye-tracking study. *Language Learning*, *68*(3), 635–668.

Cooper, N., Cutler, A., & Wales, R. (2002). Constraints of lexical stress on lexical access in English: Evidence from native and non-native listeners. *Language and Speech*, *45*(3), 207–228.

Cooper, R. M. (1974). The control of eye fixation by the meaning of spoken language: A new methodology for the real-time investigation of speech perception, memory, and language processing. *Cognitive Psychology*, *6*, 84–107.

Corps, R. E., Gambi, C., & Pickering, M. J. (2018). Coordinating utterances during turn-taking: The role of prediction, response preparation, and articulation. *Discourse Processes*, *55*(2), 230–240.

Cortina, K. S., Miller, K. F., McKenzie, R., & Epstein, A. (2015). Where low and high inference data converge: Validation of CLASS assessment of mathematics instruction using mobile eye tracking with expert and novice teachers. *International Journal of Science and Mathematics Education*, *13*(2), 389–403.

Crosby, M. E., & Stelovsky, J. (1990). How do we read algorithms? A case study. *Computer*, *23*(1), 25–35.

Crum Brown, A. (1878). Cyon's researches on the ear. *Nature*, *17*, 657–659.

Csikszentmihalyi, M. (1990). *Flow: The psychology of happiness*. Random House.

Cui, H., Liu, X. H., Wang, K. Y., Zhu, C. Y., Wang, C., & Xie, X. H. (2014). Association of saccade duration and saccade acceleration/deceleration asymmetry during visually guided saccade in schizophrenia patients. *PloS One*, *9*(5), e97308. https://doi.org/10.1371/journal.pone.0097308

Cumming, G. (2012). *Understanding the new statistics: Effect sizes, confidence intervals, and meta-analysis*. Routledge.

Curcio, C. A., Sloan, K. R., Kalina, R. E., & Hendrickson, A. E. (1990). Human photoreceptor topography. *Journal of Comparative Neurology*, *292*(4), 497–523.

Dahan, D., Tanenhaus, M. K., & Chambers, C. G. (2002). Accent and reference resolution in spoken-language comprehension. *Journal of Memory and Language*, *47*(2), 292–314.

Dale, R., & Spivey, M. J. (2018). Weaving oneself into others. In Brône, G., & Oben, B. (Eds.). *Eye-tracking in Interaction: Studies on the role of eye gaze in dialogue*, (pp. 67–90). Benjamins.

Dalrymple, K. A., Manner, M. D., Harmelink, K. A., Teska, E. P., & Elison, J. T. (2018). An examination of recording accuracy and precision from eye tracking data from toddlerhood to adulthood. *Frontiers in Psychology*, *9*(803), 1–12.

DeAngelus, M., & Pelz, J. B. (2009). Top-down control of eye movements: Yarbus revisited. *Visual Cognition*, *17*(6–7), 790–811.

De Beugher, S., Brône, G., & Goedemé, T. (2018). Automatic analysis of in-the-wild mobile eye-tracking experiments using object, face and person detection. In Brône, G., & Oben, B. (Eds.). *Eye-tracking in interaction: Studies on the role of eye gaze in dialogue*, (pp. 169–194). Benjamins.

Degen, J., & Tanenhaus, M. K. (2015). Processing scalar implicature: A constraint-based approach. *Cognitive Science*, *39*(4), 667–710.

Delabarre, E. B. (1898). A method of recording eye-movements. *The American Journal of Psychology*, *9*(4), 572–574.

De Rosis, F., Pelachaud, C., Poggi, I., Carofiglio, V., & De Carolis, B. (2003). From Greta's mind to her face: Modelling the dynamics of affective states in a conversational embodied agent. *International Journal of Human-Computer Studies*, *59*(1–2), 81–118.

De Ruiter, J. P., Mitterer, H., & Enfield, N. J. (2006). Projecting the end of a speaker's turn: A cognitive cornerstone of conversation. *Language*, *82*(3), 515–535.

Diefendorf, A. R., & Dodge, R. (1908). An experimental study of the ocular reactions of the insane from photographic records. *Brain*, *31*(3), 451–489.

Dikker, S., Wan, L., Davidesco, I., Kaggen, L., Oostrik, M., McClintock, J., . . . Poeppel, D. (2017). Brain-to-brain synchrony tracks real-world dynamic group interactions in the classroom. *Current Biology*, *27*(9), 1375–1380.

Ding, X., Yue, X., Zheng, R., Bi, C., Li, D., & Yao, G. (2019). Classifying major depression patients and healthy controls using EEG, eye tracking and galvanic skin response data. *Journal of Affective Disorders*, *251*, 156–161.

D'Mello, S. K. (2019). Gaze-based attention-aware cyberlearning technologies. In Parsons, T. D., Lin, L., & Cockerham, D. (Eds.). *Mind, brain and technology*, (pp. 87–105). Cham: Springer.

D'Mello, S. K., Mills, C., Bixler, R., & Bosch, N. (2017). Zone out no more: Mitigating mind wandering during computerized reading. International Educational Data Mining Society. 8–15.

D'Mello, S., Olney, A., Williams, C., & Hays, P. (2012). Gaze tutor: A gaze-reactive intelligent tutoring system. *International Journal of Human-Computer Studies*, *70*(5), 377–398.

Dodge, R. (1907). An experimental study of visual fixation. *The Psychological Review: Monograph Supplements*, *8*(4), i.

Doherty-Sneddon, G., Bruce, V., Bonner, L., Longbotham, S., & Doyle, C. (2002). Development of gaze aversion as disengagement from visual information. *Developmental Psychology*, *38*(3), 438–445.

Doherty-Sneddon, G., & Phelps, F. G. (2005). Gaze aversion: A response to cognitive or social difficulty?. *Memory & Cognition*, *33*(4), 727–733.

Downing, C., & Pinker, S. (1985). The spatial structure of selective attention. In Posner, M. I., & Marin, O. S. M. (Eds.). *Attention and performance XI: Mechanisms of attention and visual search*, (pp. 171–187). Routledge.

Drèze, X., & Hussherr, F. X. (2003). Internet advertising: Is anybody watching?. *Journal of Interactive Marketing*, *17*(4), 8–23.

Driver, J. (2001). A selective review of selective attention research from the past century. *British Journal of Psychology*, *92*(1), 53–78.

Duchowski, A. T. (2002). A breadth-first survey of eye-tracking applications. *Behavior Research Methods, Instruments, & Computers*, *34*(4), 455–470.

Duchowski, A. T. (2017). *Eye tracking methodology. Theory and practice* (3rd ed.). Springer.

Duchowski, A. T. (2018). Gaze-based interaction: A 30 year retrospective. *Computers & Graphics*, *73*, 59–69.

Duncan, S. (1972). Some signals and rules for taking speaking turns in conversations. *Journal of Personality and Social Psychology*, *23*(2), 283–292.

Duncan Jr, S. (1974). On the structure of speaker-auditor interaction during speaking turns. *Language in Society*, *3*(2). 161–180.

Duncan Jr, S., & Niederehe, G. (1974). On signalling that it's your turn to speak. *Journal of Experimental Social Psychology*, *10*(3), 234–247.

Dunham, P. J. & Moore, C. (1995). Current themes in research on joint attention. In Moore, C., & Dunham, P. J. (Eds.). *Joint attention: Its origin and role in development*, (pp. 15–28). Lawrence Erlbaum.

Duyck, W., Vanderelst, D., Desmet, T., & Hartsuiker, R. J. (2008). The frequency effect in second-language visual word recognition. *Psychonomic Bulletin & Review*, *15*(4), 850–855.

d'Ydewalle, G., & Gielen, I. (1992). Attention allocation with overlapping sound, image, and text. In Rayner, K. (Ed.). *Eye movements and visual cognition*, (pp. 415–427). New York: Springer.

d'Ydewalle, G., Van Rensbergen, J., & Pollet, J. (1987). Reading a message when the same message is available auditorily in another language: The case of

subtitling. In O'Regan, J. K. & Levy-Schoen, A. (Eds.). *Eye movements: From physiology to cognition*, (pp. 313–321). Amsterdam: North Holland.
Adreas Straube, A. and Büttner, U. Eggert, T. (2007). Eye movement recordings: methods. In *Neuro-Ophthalmology* (Vol. 40, pp. 15–34). Karger Publishers.
Ekman, P., & Friesen, W. V. (1978). *Facial action coding system*. Palo Alto: Consulting Psychologists Press.
ELAN (Version 6.0) [Computer software]. (2020). Nijmegen: Max Planck Institute for Psycholinguistics, The Language Archive. Retrieved from https://archive.mpi.nl/tla/elan
Emery, N. J. (2000). The eyes have it: The neuroethology, function and evolution of social gaze. *Neuroscience Biobehavior Review*, *24*(6), 581–604.
Enders, C. K. (2010). *Applied missing data analysis*. Guilford Press.
Engbert, R., & Kliegl, R. (2011). Parallel graded attention models of reading. In Liversedge, S., Gilchrist, I., & Everling, S. (Eds.). *The Oxford handbook of eye movements*, (pp. 787–800). Oxford University Press.
Engbert, R., Longtin, A., & Kliegl, R. (2002). A dynamical model of sac-cade generation in reading based on spatially distributed lexical processing. *Vision Research*, *42*(5), 621–636.
Engbert, R., Nuthmann, A., Richter, E. M., & Kliegl, R. (2005). SWIFT: a dynamical model of saccade generation during reading. *Psychological Review*, *112*(4), 777–813.
Engelhardt, P. E., Bailey, K. G., & Ferreira, F. (2006). Do speakers and listeners observe the Gricean maxim of quantity?. *Journal of Memory and Language*, *54*(4), 554–573.
Erickson, F. (2004). *Talk and social theory*. Polity Press.
Erickson, F. (2011). Uses of video in social research: a brief history. *International Journal of Social Research Methodology*, *14*(3), 179–189.
Eriksen, C. W., & St. James, J. D. (1986). Visual attention within and around the field of focal attention: A zoom lens model. *Perception & Psychophysics*, *40*, 225–240. https://doi.org/10.3758/BF03211502
Farroni, T., Csibra, G., Simion, F., & Johnson, M. H. (2002). Eye contact detection in humans from birth. *Proceedings of the National Academy of Sciences*, *99*(14), 9602–9605.
Field, A., Miles, J., & Field, Z. (2012). *Discovering Statistics Using R*. Sage.
Fodor, J. A. (1983). *The modularity of mind*. MIT Press.
Fodor, J. A. & Bever, T. G. (1965). The psychological reality of linguistic segments. *Journal of Verbal Learning and Verbal Behavior*, *4*(5), 414–420.
Fontaine, G., Cossette, S., Maheu-Cadotte, M. A., Mailhot, T., Deschênes, M. F., Mathieu-Dupuis, G., . . . & Dubé, V. (2019). Efficacy of adaptive e-learning for health professionals and students: A systematic review and meta-analysis. *BMJ Open*, *9*(8), e025252.
Forrin, N. D., Mills, C., D'Mello, S. K., Risko, E. F., Smilek, D., & Seli, P. (2021). TL; DR: longer sections of text increase rates of unintentional mind-wandering. *The Journal of Experimental Education*, 89(2), 278-290.
Foulsham, T., Farley, J., & Kingstone, A. (2013). Mind wandering in sentence reading: Decoupling the link between mind and eye. *Canadian Journal of Experimental Psychology/Revue canadienne de psychologie expérimentale*, *67*(1), 51–59.

Fox, K. C., & Beaty, R. E. (2019). Mind-wandering as creative thinking: Neural, psychological, and theoretical considerations. *Current Opinion in Behavioral Sciences*, *27*, 123–130.

Franklin, M. S., Broadway, J. M., Mrazek, M. D., Smallwood, J., & Schooler, J. W. (2013). Window to the wandering mind: Pupillometry of spontaneous thought while reading. *Quarterly Journal of Experimental Psychology*, *66*(12). 2289–2294.

Franklin, M. S., Mrazek, M. D., Anderson, C. L., Johnston, C., Smallwood, J., Kingstone, A., & Schooler, J. W. (2017). Tracking distraction: The relationship between mind-wandering, meta-awareness, and ADHD symptomatology. *Journal of Attention Disorders*, *21*(6), 475–486.

Franklin, M. S., Smallwood, J., & Schooler, J. W. (2011). Catching the mind in flight: Using behavioral indices to detect mindless reading in real time. *Psychonomic Bulletin & Review*, *18*(5), 992–997.

Frazier, L., & Rayner, K. (1982). Making and correcting errors during sentence comprehension: Eye movements in the analysis of structurally ambiguous sentences. *Cognitive Psychology*, *14*(2), 178–210.

Friederici, A. D. (2011). The brain basis of language processing: from structure to function. *Physiological Reviews*, *91*(4), 1357–1392.

Friedman, A. (1979). Framing pictures: The role of knowledge in automatized encoding and memory for gist. *Journal of Experimental Psychology: General*, *108*(3), 316–355.

Friesen, C. K., & Kingstone, A. (1998). The eyes have it! Reflexive orienting is triggered by nonpredictive gaze. *Psychonomic Bulletin & Review*, *5*(3), 490–495.

Frisson, S., Rayner, K., & Pickering, M. J. (2005). Effects of contextual predictability and transitional probability on eye movements during reading. *Journal of Experimental Psychology: Learning, Memory, and Cognition*, *31*(5), 862–877.

Galéra, C., Orriols, L., M'Bailara, K., Laborey, M., Contrand, B., Ribéreau-Gayon, R., . . . & Maury, B. (2012). Mind wandering and driving: responsibility case-control study. *BMJ (British Medical Journal)*, *345*, e8105.

Garfinkel, H. (1967). *Studies in ethnomethodology*. Englewood cliffs: Prentice-Hall.

Gass, S., Winke, P., Isbell, D. R., & Ahn, J. (2019). How captions help people learn languages: A working-memory, eye-tracking study. *Language Learning & Technology*, *23*(2), 84–104. https://doi.org/10125/44684

Gegenfurtner, A., Lehtinen, E., & Säljö, R. (2011). Expertise differences in the comprehension of visualizations: A meta-analysis of eye-tracking research in professional domains. *Educational Psychology Review*, *23*(4), 523–552.

Giacomantonio, M., Jordan, J., Federico, F., van den Assem, M. J., & van Dolder, D. (2018). The evil eye: Eye gaze and competitiveness in social decision making. *European Journal of Social Psychology*, *48*(3), 388–396.

Gilzenrat, M. S., Nieuwenhuis, S., Jepma, M., & Cohen, J. D. (2010). Pupil diameter tracks changes in control state predicted by the adaptive gain theory of locus coeruleus function. *Cognitive, Affective, & Behavioral Neuroscience*, *10*(2), 252–269.

Gironzetti, E. (2017). Multimodal and eye-tracking evidence in the negotiation of pragmatic intentions in dyadic conversations: the case of humorous dis-course. Unpublished PhD dissertation. Texas A&M University-Commerce.

Gironzetti, E., Attardo, S., & Pickering, L. (2016a). Smiling, gaze and humor in conversation. In Ruiz-Gurillo, E. (Ed.). *Metapragmatics of humor: Current Research Trends* (pp. 237–256). Amsterdam: Benjamins.

Gironzetti, E., Pickering, L., Huang, M., Zhang, Y., Menjo, S., & Attardo, S. (2016b). Smiling synchronicity and gaze patterns in dyadic humorous conversations. *Humor: International Journal of Humor Research*, *29*(2), 301–324.

Gleitman, L. R., January, D., Nappa, R., & Trueswell, J. C. (2007). On the give and take between event apprehension and utterance formulation. *Journal of Memory and Language*, *57*(4), 544–569.

Glenberg, A. M., Schroeder, J. L., & Robertson, D. A. (1998). Averting the gaze disengages the environment and facilitates remembering. *Memory & Cognition*, *26*(4), 651–658.

Gluck, K. A., Anderson, J. R., & Douglass, S. A. (2000). Broader bandwidth in student modeling: What if ITS Were "Eye" TS?. In International Conference on Intelligent Tutoring Systems (pp. 504–513). Berlin, Heidelberg: Springer.

Godfroid, A. (2020). *Eye tracking in second language acquisition and bilingual-ism: A research synthesis and methodological guide*. Routledge.

Godfroid, A., Boers, F., & Housen, A. (2013). An eye for words: Gauging the role of attention in incidental L2 vocabulary acquisition by means of eye-tracking. *Studies in Second Language Acquisition*, *35*(3), 483–517.

Godfroid, A., Housen, A., & Boers, F. (2010). A procedure for testing the Noticing Hypothesis in the context of vocabulary acquisition. In Pütz, M. & Sicola, L.(Eds.). *Cognitive processing in second language acquisition*, (pp. 169–197). Benjamins.

Godfroid, A., Winke, P., & Conklin, K. (2020). Exploring the depths of second language processing with eye tracking: An introduction. *Second Language Research*, *36*(3), 243–255.

Goodwin, C. (1981). *Conversational organization. Interactions between speakers and hearers*. Academic Press.

Goujon, A., Brockmole, J. R., & Ehinger, K. A. (2012). How visual and semantic information influence learning in familiar contexts. *Journal of Experimental Psychology: Human Perception and Performance*, *38*(5), 1315–1327.

Grandchamp, R., Braboszcz, C., & Delorme, A. (2014). Oculometric variations during mind wandering. *Frontiers in Psychology*, *5*, 31.

Grant, M., Sloggett, S., & Dillon, B. (2020). Processing ambiguities in attachment and pronominal reference. *Glossa: A Journal of General Linguistics*, *5*(1), 1–30.

Grice, H. P. (1989). *Studies in the way of words*. Harvard University Press.

Gries, S. Th. (2013). *Statistics for linguistics with R*. Mouton De Gruyter.

Griffin, Z. M., (2004). Why look? Reasons for eye movements related to language production. In Henderson, J., & Ferreira, F. (Eds.). *The interface of language, vision, and action: Eye movements and the visual world*, (pp. 192–222). Psychology Press.

Griffin, Z. M., & Bock, K. (2000). What the eyes say about speaking. *Psychological Science*, *11*(4), 274–279.

Grodner, D., & Sedivy, J. C. (2011). The effect of speaker-specific information on pragmatic inferences. In Gibson, E., & Pearlmutter, N. J. (Eds.). *The processing and acquisition of reference*. (pp. 239–272). MIT Press.

Grossmann, T. (2017). The eyes as windows into other minds: An integrative perspective. *Perspectives on Psychological Science*, *12*(1), 107–121.

Grüter, T., Lau, E., & Ling, W. (2018). L2 listeners rely on the semantics of classifiers to predict. In Proceedings of the 42nd Annual Boston University Conference on Language Development (pp. 303–316).

Grüter, T., Rohde, H., & Schafer, A. (2014). The role of discourse-level expectations in non-native speakers' referential choices. In Orman, W. & Valleau, M. J. (Eds.). Proceedings of the Annual Boston University Conference on Language Development. (Vol. 1, pp. 179–191). Somerville: Cascadilla Press.

Gu, Y., Wang, C., Bixler, R., & D'Mello, S. (2017). ETGraph: A graph-based approach for visual analytics of eye-tracking data. *Computers & Graphics*, *62*, 1–14.

Gullberg, M., & Holmqvist, K. (1999). Keeping an eye on gestures: Visual perception of gestures in face-to-face communication. *Pragmatics & Cognition*, *7*(1), 35–63.

Gullberg, M., & Holmqvist, K. (2006). What speakers do and what addressees look at: Visual attention to gestures in human interaction live and on video. *Pragmatics & Cognition*, *14*(1), 53–82.

Gullberg, M., & Kita, S. (2009). Attention to speech-accompanying gestures: Eye movements and information uptake. *Journal of Nonverbal Behavior*, *33*(4), 251–277.

Gumperz, J. J. (2015). Interactional sociolinguistics: a personal perspective. In Tannen, D., Hamilton, H. E., & Schiffrin, D. (Eds.). *The handbook of discourse analysis*, (pp. 309–323). Chichester: John Wiley & Sons.

Gunner, J. (1998). Iconic discourse: The troubling legacy of Mina Shaughnessy. *Journal of Basic Writing*, *17*(2). 25–42.

Haataja, E., Salonen, V., Laine, A., Toivanen, M., & Hannula, M. S. (2020). The relation between teacher-student eye contact and teachers' interpersonal behavior during group work: A multiple-person gaze-tracking case study in secondary mathematics education. *Educational Psychology Review*, *33*(1), 1–17.

Haensel, J. X., Danvers, M., Ishikawa, M., Itakura, S., Tucciarelli, R., Smith, T. J., & Senju, A. (2020). Culture modulates face scanning during dyadic social interactions. *Scientific Reports*, *10*(1), 1–11.

Hall, C. A., & Chilcott, R. P. (2018). Eyeing up the future of the pupillary light reflex in neurodiagnostics. *Diagnostics*, *8*(1), 19. https://doi.org/10.3390/diagnostics8010019

Hanna, J. E., & Brennan, S. E. (2007). Speakers' eye gaze disambiguates referring expressions early during face-to-face conversation. *Journal of Memory and Language*, *57*(4), 596–615.

Hanna, J. E., Tanenhaus, M. K., & Trueswell, J. C. (2003). The effects of common ground and perspective on domains of referential interpretation. *Journal of Memory and Language*, *49*(1), 43–61.

Harezlak, K., & Kasprowski, P. (2018). Application of eye tracking in medicine: A survey, research issues and challenges. *Computerized Medical Imaging and Graphics*, *65*, 176–190.

Hartridge, H. & Thompson, L. C. (1948). Methods of investigating eye movements. *British Journal of Ophthalmology*, *32*, 581–591.

Hatcher, L. (2013). *Advanced statistics in research: Reading, understanding, and writing up data analysis results*. ShadowFinch Media, LLC.

Hayakawa, S., & Marian, V. (2020). Studying bilingualism through eye-tracking and brain imaging. In Heredia, R. R., & Cieślicka, A. B. (Eds.). *Bilingual lexical ambiguity resolution*, (pp. 273–299). Cambridge.

He, J., Becic, E., Lee, Y. C., & McCarley, J. S. (2011). Mind wandering behind the wheel: performance and oculomotor correlates. *Human Factors*, *53*(1), 13–21.

He, J., Siu, C., Chaparro, B., & Strohl, J. (2014). Mobile. In Bergstrom, J. R. & Schall, A. J. (Eds.). *Eye tracking in user experience design*, (pp. 255–290). Elsevier.

Heigham, J. & Croker, R. A. (2009). *Qualitative research in applied linguistics*. London: Palgrave.

Henderson, J. M., & Hollingworth, A. (1999). High-level scene perception. *Annual Review of Psychology*, *50*(1), 243–271.

Hermena, E. W., Drieghe, D., Hellmuth, S., & Liversedge, S. P. (2015). Processing of Arabic diacritical marks: Phonological–syntactic disambiguation of homographic verbs and visual crowding effects. *Journal of Experimental Psychology: Human Perception and Performance*, *41*(2), 494–507.

Hervet, G., Guérard, K., Tremblay, S., & Chtourou, M. S. (2010). Is banner blindness genuine? Eye tracking internet text advertising. *Applied Cognitive Psychology*, *25*(5), 708–716.

Hessels, R. S. (2020). How does gaze to faces support face-to-face interaction? A review and perspective. *Psychonomic Bulletin & Review*, 27, 856–881.

Hessels, R. S., Cornelissen, T. H. W., Hooge, I. T. C., & Kemner, C. (2017). Gaze behavior to faces during dyadic interaction. *Canadian Journal of Experimental Psychology-Revue Canadienne De Psychologie Experimentale*, *71*(3), 226–242. https ://doi.org/10.1037/cep00 00113.

Higgins, E., Leinenger, M., & Rayner, K. (2014). Eye movements when viewing advertisements. *Frontiers in Psychology, 5*(210), 1–15.

Hills, P. J., Cooper, R. E., & Pake, J. M. (2013). Removing the own-race bias in face recognition by attentional shift using fixation crosses to diagnostic features: An eye-tracking study. *Visual Cognition*, *21*(7), 876–898.

Hills, P. J., & Pake, J. M. (2013). Eye-tracking the own-race bias in face recognition: Revealing the perceptual and socio-cognitive mechanisms. *Cognition*, *129*(3), 586–597.

Hirotani, M., Frazier, L., & Rayner, K. (2006). Punctuation and intonation effects on clause and sentence wrap-up: Evidence from eye movements. *Journal of Memory and Language*, *54*(3), 425–443.

Ho, S., Foulsham, T., & Kingstone, A. (2015). Speaking and listening with the eyes: gaze signaling during dyadic interactions. *PloS One*, *10*(8), e0136905, 1-18.

Holler, J., & Kendrick, K. H. (2015). Unaddressed participants' gaze in multi-person interaction: optimizing recipiency. *Frontiers in Psychology, 6*(98), 1–14.

Hollis, R. B., & Was, C. A. (2016). Mind wandering, control failures, and social media distractions in online learning. *Learning and Instruction*, *42*, 104–112.

Holmqvist, K., & Andersson, R. (2017). *Eye tracking: A comprehensive guide to methods, paradigms and measures*. Lund Eye-Tracking Research Institute.

Holsanova, J. (2014). Reception of multimodality: Applying eye tracking methodology in multimodal research. In: Jewitt, C. (Ed.). *Routledge handbook of multimodal analysis* (2nd ed. Chapter 20, pp. 285–296). Routledge.

Holsanova, J., Holmberg, N., & Holmqvist, K. (2009). Reading information graphics: The role of spatial contiguity and dual attentional guidance. *Applied Cognitive Psychology: The Official Journal of the Society for Applied Research in Memory and Cognition*, *23*(9), 1215–1226.

Hornof, A., Cavender, A., & Hoselton, R. (2003). Eyedraw: a system for drawing pictures with eye movements. *ACM SIGACCESS Accessibility and Computing*, *77–78*, 86–93.

Huang, Y. T., & Snedeker, J. (2009). Online interpretation of scalar quantifiers: Insight into the semantics-pragmatics interface. *Cognitive Psychology*, *58*(3), 376–415.

Hubel, D. H. (1995). *Eye, Brain and Vision*. New York: Scientific American.

Huckauf, A., & Urbina, M. H. (2008). Gazing with pEYEs: towards a universal input for various applications. In Proceedings of the 2008 Symposium on Eye Tracking Research & Applications (pp. 51–54).

Huddleston, P., Behe, B. K., Minahan, S., & Fernandez, R. T. (2015). Seeking attention: An eye tracking study of in-store merchandise displays. *International Journal of Retail and Distribution Management*, *43*(6), 561–574.

Huettig, F. (2015). Four central questions about prediction in language processing. *Brain Research*, *1626*, 118–135.

Huettig, F., & Guerra, E. (2019). Effects of speech rate, preview time of visual context, and participant instructions reveal strong limits on prediction in language processing. *Brain Research*, 1706, 196–208.

Huettig, F., & Janse, E. (2016). Individual differences in working memory and processing speed predict anticipatory spoken language processing in the visual world. *Language, Cognition and Neuroscience*, *31*(1), 80–93.

Huettig, F., Rommers, J., & Meyer, A. S. (2011). Using the visual world paradigm to study language processing: A review and critical evaluation. *Acta Psychologica*, *137*(2), 151–171.

Huey, E. B. (1898). Preliminary experiments in the physiology and psychology of reading. *The American Journal of Psychology*, *9*(4), 575–586.

Huey, E. B. (1908). *The psychology and pedagogy of reading*. Macmillan.

Hutchinson, T. E., White, K. P., Martin, W. N., Reichert, K. C., & Frey, L. A. (1989). Human-computer interaction using eye-gaze input. *IEEE Transactions on Systems, Man, and Cybernetics*, *19*(6), 1527–1534.

Hutt, S., Krasich, K., Mills, C., Bosch, N., White, S., Brockmole, J. R., & D'Mello, S. K. (2019). Automated gaze-based mind wandering detection during computerized learning in classrooms. *User Modeling and User-Adapted Interaction*, *29*(4), 821–867.

Hutt, S., Mills, C., Bosch, N., Krasich, K., Brockmole, J., & D'mello, S. (2017). "Out of the Fr-Eyeing Pan" towards Gaze-based models of attention during learning with technology in the classroom. In Proceedings of the 25th Conference on User Modeling, Adaptation and Personalization (pp. 94–103).

Hutt, S., Mills, C., White, S., Donnelly, P. J., & D'Mello, S. K. (2016). The eyes have it: Gaze-based detection of mind wandering during learning with an intelligent tutoring system. Proceedings of the 9th International Conference on Educational Data Mining (EDM) (pp. 86–93). International Educational Data Mining Society.

Hymes, D. (1972). Models of the interaction of language and social life. In J. J., Gumperz, & D. Hymes (Eds.). *Directions in sociolinguistics: The ethnography of communication*, (pp. 35–71). New York: Holt, Rinehart & Winston.

Hyönä, J., & Kaakinen, J. K. (2019). Eye movements during reading. In C. Klein & U. Ettinger (eds), *Eye Movement Research* (pp. 239–274). Cham: Springer.

Hyönä, J., Pollatsek, A., Koski, M., & Olkoniemi, H. (2020). An eye-tracking study of reading long and short novel and lexicalized compound words. *Journal of Eye Movement Research*, *13*(4). 1–17.

Itkonen, T., Pekkanen, J., & Lappi, O. (2015). Driver gaze behavior is different in normal curve driving and when looking at the tangent point. *PloS One*, 10(8), e0135505.

Ito, A., Pickering, M. J., & Corley, M. (2018). Investigating the time-course of phonological prediction in native and non-native speakers of English: A visual world eye-tracking study. *Journal of Memory and Language*, *98*, 1–11.

Itti, L., & Koch, C. (2000). A saliency-based search mechanism for overt and covert shifts of visual attention. *Vision Research*, 40(10–12), 1489–1506.

Jacob, R. J. K. (1991). The use of eye movements in human-computer interaction techniques: what you look at is what you get. *ACM Transactions on Information Systems (TOIS)*, *9*(2), 152–169.

Jacob, R. J. K. & Karn, K. S. (2003). Eye tracking in human-computer interaction and usability research: Ready to deliver the promises. In Hyönä, J., Radach, R. and Deubel, H. (Eds.). *The mind's eye cognitive and applied aspects of eye movement research*, (pp. 573–605). Elsevier.

Jang, D., Yang, I., & Kim, S. (2020). Detecting mind-wandering from eye movement and oculomotor data during learning video lecture. *Education Sciences*, *10*(3), 51.

Janisse, M. P. (1977). *Pupillometry. The psychology of the pupillary response*. New York: Wiley.

Jarick, M., & Kingstone, A. (2015). The duality of gaze: Eyes extract and signal social information during sustained cooperative and competitive dyadic gaze. *Frontiers in Psychology*, *6*(1423), 1–7.

Jarodzka, H., Skuballa, I., & Gruber, H. (2020). Eye-tracking in educational practice: Investigating visual perception underlying teaching and learning in the classroom. *Educational Psychology Review*, 1–10.

Jermann, P., Nüssli, M. A., & Li, W. (2010). Using dual eye-tracking to unveil coordination and expertise in collaborative Tetris. In McEwan, T. & McKinnon, L. M. (Eds.). *People and Computer. XXIV Games are a Serious Business.* Proceedings of the 24th BCS Interaction Specialist Group Conference. (pp. 36–44). Swindon: British Informatics Society.

Jewitt, C. E. (2014). *The Routledge handbook of multimodal analysis*. (2nd ed.). Routledge/Taylor & Francis Group.

Johnson, C. I., & Mayer, R. E. (2012). An eye movement analysis of the spatial contiguity effect in multimedia learning. *Journal of Experimental Psychology: Applied*, *18*(2), 178.

Johnson, K. (2008). *Quantitative methods in linguistics*. Blackwell.

Jokinen, K., Furukawa, H., Nishida, M., & Yamamoto, S. (2013). Gaze and turn-taking behavior in casual conversational interactions. *ACM Transactions on Interactive Intelligent Systems (TiiS)*, *3*(2), 1–30.

Jongerius, C., Hessels, R. S., Romijn, J. A., Smets, E., & Hillen, M. A. (2020). The measurement of eye contact in human interactions: A scoping review. *Journal of Nonverbal Behavior*, *44*, 363–389.

Judd, C. H. (1907). Photographic records of convergence and divergence. *The Psychological Review: Monograph Supplements*, *8*(3), 370–423.

Judd, C. H., McAllister, C. N., & Steele, W. M. (1905). General introduction to a series of studies of eye movements by means of kinetoscopic photographs. *Psychological Review Monographs*, *7*(1), 1–16.
Juhasz, B. J., & Sheridan, H. (2020). The time course of age-of-acquisition effects on eye movements during reading: Evidence from survival analyses. *Memory & Cognition*, *48*(1), 83–95.
Just, M. A., & Carpenter, P. A. (1980). A theory of reading: From eye fixations to comprehension. *Psychological Review*, *87*(4), 329–354.
Kaan, E. (2014). Predictive sentence processing in L2 and L1: What is different? *Linguistic Approaches to Bilingualism*, *4*(257), 282.
Kaan, E., Dallas, A. C., & Wijnen, F. (2010). Syntactic predictions in second-language sentence processing. In Zwart, J.-W. & de Vries, M. (Eds.). *Structure preserved. Festschrift in the honor of Jan Koster* (pp. 207–213). Benjamins.
Kachigan, S. K. (1982). *Statistical analysis. An interdisciplinary introduction to univariate and multivariate methods*. New York: Radius Press.
Kamide, Y., Scheepers, C., & Altmann, G. T. (2003). Integration of syntactic and semantic information in predictive processing: Cross-linguistic evidence from German and English. *Journal of Psycholinguistic Research*, *32*(1), 37–55.
Kane, M. J., Smeekens, B. A., Von Bastian, C. C., Lurquin, J. H., Carruth, N. P., Miyake, A. (2017). A combined experimental and individual-differences investigation into mind wandering during a video lecture. *Journal of Experimental Psychology: General*, *146*(11), 1649–1674.
Karpov, B. A., Luria, A. R., & Yarbus, A. L. (1968). Disturbances of the structure of active perception in lesions of the posterior and anterior regions of the brain. *Neuropsychologia*, *6*(2), 157–166.
Kabashima, K., Jokinen, K., Nishida, M., & Yamamoto, S. (2012). Multimodal corpus of conversations in mother tongue and second language by same interlocutors. In Proceedings of the 4th Workshop on Eye Gaze in Intelligent Human Machine Interaction (pp. 1–5).
Kelly, D. J., Miellet, S., & Caldara, R. (2010). Culture shapes eye movements for visually homogeneous objects. *Frontiers in Psychology, 1*(6), 1-7.
Kendon, A. (1967). Some functions of gaze-direction in social interaction. *Acta Psychologica*, *26*, 22–63.
Keysar, B., Barr, D. J., Balin, J. A., & Brauner, J. S. (2000). Taking perspective in conversation: The role of mutual knowledge in comprehension. *Psychological Science*, *11*(1), 32–38.
Keysar, B., Barr, D. J., & Horton, W. S. (1998). The egocentric basis of language use: Insights from a processing approach. *Current Directions in Psychological Science*, *7*(2), 46–49.
Kidwell, M. (2009). Gaze shift as an interactional resource for very young children. *Discourse Processes*, *46*(2–3), 145–160.
Kidwell, M. (2014). The role of gaze in conversational interaction. In A. Cienki, E. Fricke, D. McNeill, & C. Müller (Eds.). *Body - Language - Communication; Linguistics and Communication Sciences: An International Handbook*. (pp. 1324–1333). Berlin: Mouton de Gruyter.
Kidwell, M. (2015). Gaze. In K. Tracy (Ed.). *The international encyclopedia of language and social interaction*. (pp. 657–661). Wiley.
Killingsworth, M. A., & Gilbert, D. T. (2010). A wandering mind is an unhappy mind. *Science*, *330*(6006), 932–932.

Kim, J., Guo, P. J., Seaton, D. T., Mitros, P., Gajos, K. Z., & Miller, R. C. (2014). Understanding in-video dropouts and interaction peaks in online lecture videos. In Proceedings of the First ACM Conference on Learning Scale Conference (pp. 31–40).

Kim, K., & Mundy, P. (2012). Joint attention, social-cognition, and recognition memory in adults. *Frontiers in Human Neuroscience, 6*(172), 1–11.

Kinreich, S., Djalovski, A., Kraus, L., Louzoun, Y. & Feldman, R. (2017). Brain-to-Brain Synchrony during Naturalistic Social Interactions. *Scientific Reports, 7*(17060), 1–12. https://doi.org/10.1038/s41598-017-17339-5

Kirschner, P. A. (2017). Stop propagating the learning styles myth. *Computers & Education*, *106*, 166–171.

Kleinke, C. L. (1986). Gaze and eye contact—A research review. *Psychological Bulletin*, *100*(1), 78–100. https ://doi.org/10.1037/0033-2909.100.1.78.

Klingner, J., Kumar, R., & Hanrahan, P. (2008). Measuring the task-evoked pupillary response with a remote eye tracker. In Proceedings of the 2008 Symposium on Eye Tracking Research & Applications (pp. 69–72). Association for Computing Machinery.

Komogortsev, O. V., Gobert, D. V., Jayarathna, S., & Gowda, S. M. (2010). Standardization of automated analyses of oculomotor fixation and saccadic behaviors. *IEEE Transactions on Biomedical Engineering*, *57*(11), 2635–2645.

Knoeferle, P., Crocker, M. W., Scheepers, C., & Pickering, M. J. (2005). The influence of the immediate visual context on incremental thematic role-assignment: Evidence from eye-movements in depicted events. *Cognition*, *95*(1), 95–127.

Knoeferle, P., Pyykkönen-Klauck, P. & Crocker, M. W. (Eds.). (2016). *Visually Situated Language Comprehension*. Benjamins.

Ko, H. K., Snodderly, D. M., & Poletti, M. (2016). Eye movements between saccades: Measuring ocular drift and tremor. *Vision Research*, *122*, 93–104.

Koć-Januchta, M., Höffler, T., Thoma, G. B., Prechtl, H., & Leutner, D. (2017). Visualizers versus verbalizers: Effects of cognitive style on learning with texts and pictures–An eye-tracking study. *Computers in Human Behavior*, *68*, 170–179.

Konopka, A. E. (2019). Encoding actions and verbs: Tracking the time-course of relational encoding during message and sentence formulation. *Journal of Experimental Psychology: Learning, Memory, and Cognition*, *45*(8), 1486–1510.

Konopka, A. E., & Brown-Schmidt, S. (2014). Message encoding. In Goldrick, M. A., Ferreira, V. S., & Miozzo, M. (Eds.). *The Oxford handbook of language production*, (pp. 3–20). Oxford.

Kopp, S., & Bergmann, K. (2013). Automatic and strategic alignment of co-verbal gestures in dialogue. In Wachsmuth, I., De Ruiter, J., Jaecks, P., & Kopp, S. (Eds.). *Alignment in communication: Towards a new theory of communication*, (pp. 87–107). Benjamins.

Kousaie, S., & Phillips, N. A. (2012). Ageing and bilingualism: Absence of a "bilingual advantage" in Stroop interference in a nonimmigrant sample. *Quarterly Journal of Experimental Psychology*, *65*(2), 356–369.

Krafka, K., Khosla, A., Kellnhofer, P., Kannan, H., Bhandarkar, S., Matusik, W., & Torralba, A. (2016). Eye tracking for everyone. In Proceedings of the IEEE Conference on Computer Vision and Pattern Recognition (pp. 2176–2184).

Krasich, K., McManus, R., Hutt, S., Faber, M., D'Mello, S. K., & Brockmole, J. R. (2018). Gaze-based signatures of mind wandering during real-world scene processing. *Journal of Experimental Psychology: General*, *147*(8), 1111–1124.

Kreysa, H., Nunnemann, E. M., & Knoeferle, P. (2018). Distinct effects of different visual cues on sentence comprehension and later recall: The case of speaker gaze versus depicted actions. *Acta Psychologica*, *188*, 220–229.

Kreysa, H., & Pickering, M. J. (2011). In eye movements in dialogue. In Liversedge, S. P., Gilchrist, I. D. & Everling, S. (Eds.). *The Oxford handbook of eye movements*, (pp. 943–959). Oxford University Press.

Kress, G. & Van Leeuwen, T. (2001). *Multimodal discourse. The modes and media of contemporary communication*. Hachette.

Krueger, E., Schneider, A., Sawyer, B., Chavaillaz, A., Sonderegger, A., Groner, R., & Hancock, P. (2019). Microsaccades distinguish looking from seeing. *Journal of Eye Movement Research*, *12*(6). https://doi.org/10.16910/jemr.12.6.2

Kütt, G. H., Lee, K., Hardacre, E., & Papoutsaki, A. (2019). Eye-write: Gaze sharing for collaborative writing. In Proceedings of the 2019 CHI Conference on Human Factors in Computing Systems (pp. 1–12). New York: Association for Computing Machinery.

Ladeira, W. J., Nardi, V. A. M., Santini, F. D. O., & Jardim, W. C. (2019). Factors influencing visual attention: A meta-analysis. *Journal of Marketing Management*, *35*(17–18), 1710–1740.

Laeng, B., & Alnaes, D. (2019). Pupillometry. In Klein, C. & Ettinger, U. (Eds.). *Eye Movement Research*, (pp. 449–502). Cham: Springer.

Laeng, B., Ørbo, M., Holmlund, T., & Miozzo, M. (2011). Pupillary stroop effects. *Cognitive Processing*, *12*(1), 13–21.

Lai, M. L., Tsai, M. J., Yang, F. Y., Hsu, C. Y., Liu, T. C., Lee, S. W. Y., . . . & Tsai, C. C. (2013). A review of using eye-tracking technology in exploring learning from 2000 to 2012.*Educational Research Review*, *10*, 90–115.

Lallé, S., Toker, D., & Conati, C. (2019). Gaze-driven adaptive interventions for magazine-style narrative visualizations. IEEE Transactions on Visualization and Computer Graphics.

Lamare, M. (1892). Des mouvements des yeux dans la lecture. *Bulletins et Mémoires de la Société Française d'Ophthalmologie*, *10*, 354–364.

Lambert, R. H., Monty, R. A., & Hall, R. J. (1974). High-speed data processing and unobtrusive monitoring of eye movements. *Behavior Research Methods & Instrumentation*, *6*(6), 525–530.

Land, M., & Lee, D. (1994). Where we look when we steer. *Nature*, *369*(6483), 742–744.

Langner, M., Toreini, P., & Maedche, A. (2020). AttentionBoard: A quantified-self dashboard for enhancing attention management with eye-tracking. In Davis, F. D., Riedl, R. vom Brocke, J., Léger, P. M., Randolph, A. B., & Fischer, Y. (Eds.). *Information systems and neuroscience. NeuroIS Retreat 2020* (pp. 266–275). Cham: Springer.

Lappi, O., Pekkanen, J., Rinkkala, P., Tuhkanen, S., Tuononen, A., & Virtanen, J. P. (2020). Humans use optokinetic eye movements to track waypoints for steering. *Nature Scientific Reports*, *10*(1), 1–14.

Larsson, M. (2013). The optic chiasm: a turning point in the evolution of eye/hand coordination. *Frontiers in Zoology*, *10*(1), 1–12.

Laurence, P. (1993). The vanishing site of Mina Shaughnessy's "errors and expectations". *Journal of Basic Writing*, *12*(2), 18–28.

Lee, S. K. (2007). Effects of textual enhancement and topic familiarity on Korean EFL students' reading comprehension and learning of passive form. *Language Learning*, *57*(1), 87–118.

Legge, G. E., & Bigelow, C. A. (2011). Does print size matter for reading? A review of findings from vision science and typography. *Journal of Vision*, *11*(5), 1–22.

Legge, G. E., Klitz, T. S., & Tjan, B. S. (1997). Mr. Chips: An ideal-observer model of reading. *Psychological Review*, *104*(3), 524–553.

Legge, G. E., Pelli, D. G., Rubin, G. S., & Schleske, M. M. (1985). Psychophysics of reading? 1. Normal vision. *Vision Research*, *25*(2), 239–252.

Levine, P. & Scollon, R. (Eds.) (2004). *Discourse and technology. Multimodal discourse analysis*. Georgetown University Press.

Levinson, S. C., & Torreira, F. (2015). Timing in turn-taking and its implications for processing models of language. *Frontiers in Psychology*, *6*(731), 1–17.

Lew, R. (2015). Research into the use of online dictionaries. *International Journal of Lexicography*, *28*(2), 232–253.

Lew, R., & De Schryver, G. M. (2014). Dictionary users in the digital revolution. *International Journal of Lexicography*, *27*(4), 341–359.

Lew, R., Grzelak, M., & Leszkowicz, M. (2013). How dictionary users choose senses in bilingual dictionary entries: An eye-tracking study. *Lexikos*, *23*, 228–254.

Li, I., Dey, A., & Forlizzi, J. (2010). A stage-based model of personal informatics systems. In Proceedings of the SIGCHI Conference on Human Factors in Computing Systems (pp. 557–566). New York: Association for Computing Machinery.

Li, J., Antonenko, P. D., & Wang, J. (2019). Trends and issues in multimedia learning research in 1996–2016: A bibliometric analysis. *Educational Research Review*, 28, 100282.

Libby, W. L., & Yaklevich, D. (1973). Personality determinants of eye contact and direction of gaze aversion. *Journal of Personality and Social Psychology*, *27*(2), 197–206.

Liechty, J., Pieters, R., & Wedel, M. (2003). Global and local covert visual attention: Evidence from a Bayesian hidden Markov model. *Psychometrika*, *68*(4), 519–541.

Lim, J. Z., Mountstephens, J., & Teo, J. (2020). Emotion recognition using eye-tracking: Taxonomy, review and current challenges. *Sensors*, *20*(8), 2384, 1–21.

Liu, H. C. & Chuang, H. H. (2011). An examination of cognitive processing of multimedia information based on viewers' eye movements. *Interactive Learning Environments*, *19*(5). 503–517.

Liu, Y., Reichle, E. D., & Gao, D. G. (2013). Using reinforcement learning to examine dynamic attention allocation during reading. *Cognitive Science*, *37*(8), 1507–1540.

Liversedge, S.P., Gilchrist, I., & Everling, S. (Eds.). (2011). *The Oxford handbook of eye movements*. Oxford University Press.

Liversedge, S. P., Rayner, K., White, S. J., Vergilino-Perez, D., Findlay, J. M., & Kentridge, R. W. (2004). Eye movements when reading disappearing text: Is there a gap effect in reading?. *Vision Research*, *44*(10), 1013–1024.

Luo, Z., O'Steen, B., & Brown, C. (2020). The use of eye-tracking technology to identify visualisers and verbalisers: Accuracy and contributing factors. Interactive Technology and Smart Education.

Macdonald, R. G., & Tatler, B. W. (2018). Gaze in a real-world social interaction: A dual eye-tracking study. *Quarterly Journal of Experimental Psychology*, *71*(10), 2162–2173.

Mackworth, N. H., & Thomas, E. L. (1962). Head-mounted eye-marker camera. *Journal of the Optical Society of America*, *52*(6), 713–716.

MacLeod, C. M. (1991). Half a century of research on the Stroop effect: An integrative review. *Psychological Bulletin*, *109*(2), 163–203.

Magnuson, J. S. (2019). Fixations in the visual world paradigm: Where, when, why?. *Journal of Cultural Cognitive Science*, *3*, 113–139.

Majaranta, P., & Bulling, A. (2014). Eye tracking and eye-based human-computer interaction. In Fairclough, S. H. & Gilleade, K. (Eds.). *Advances in physiological computing* (pp. 39–65). London: Springer.

Majaranta, P., & Räihä, K. J. (2002). Twenty years of eye typing: Systems and design issues. In Proceedings of the 2002 Symposium on Eye Tracking Research & Applications (pp. 15–22).

Marsden, E., Thompson, S., & Plonsky, L. (2017). Open science in second language acquisition research: The IRIS repository of research materials and data. In SHS Web of Conferences (Vol. 38, p. 00013). EDP Sciences.

Marslen-Wilson, W. D. (1987). Functional parallelism in spoken word-recognition. *Cognition*, *25*(1–2), 71–102.

Marslen-Wilson, W., & Tyler, L. K. (1980). The temporal structure of spoken language understanding. *Cognition*, *8*(1), 1–71.

Martinez-Conde, S., & Macknik, S. L. (2007). Windows on the mind. *Scientific American*, *297*(2), 56–63.

Martinez-Conde, S., & Macknik, S. L. (2011). Microsaccades. In Liversedge, S. P., Gilchrist, I. D., & Everling, S. (Eds.). *The Oxford handbook of eye movements*, (pp. 95–114). Oxford University Press.

Martinez-Conde, S., & Macknik, S. L. (2017). Unchanging visions: The effects and limitations of ocular stillness. *Philosophical Transactions of the Royal Society B: Biological Sciences*, *372*(1718), 20160204.

Martinez-Conde, S., Macknik, S. L., & Hubel, D. H. (2004). The role of fixational eye movements in visual perception. *Nature Reviews. Neuroscience*, *5*(3), 229–240. https://doi.org/10.1038/nrn1348

Mayer, R. E. (2010). Unique contributions of eye-tracking research to the study of learning with graphics. *Learning and Instruction*, *20*(2), 167–171.

McCamy, M. B., Otero-Millan, J., Macknik, S. L., Yang, Y., Troncoso, X. G., Baer, S. M., Crook, S. M. & Martinez-Conde, S. (2012). Microsaccadic efficacy and contribution to foveal and peripheral vision. *Journal of Neuroscience*, *32*(27), 9194–9204.

McConkie, G. W., & Rayner, K. (1975). The span of the effective stimulus during a fixation in reading. *Perception & Psychophysics*, *17*(6), 578–586.

McDonald, S. A., & Shillcock, R. C. (2003). Low-level predictive inference in reading: The influence of transitional probabilities on eye movements. *Vision Research*, *43*(16), 1735–1751.

McGurk, H., & MacDonald, J. (1976). Hearing lips and seeing voices. *Nature*, *264*(5588), 746–748.

McIntyre, N. A., & Foulsham, T. (2018). Scanpath analysis of expertise and culture in teacher gaze in real-world classrooms. *Instructional Science*, *46*(3), 435–455.

McIntyre, N. A., Mainhard, M. T., & Klassen, R. M. (2017). Are you looking to teach? Cultural, temporal and dynamic insights into expert teacher gaze. *Learning and Instruction*, *49*, 41–53.

McVay, J. C., & Kane, M. J. (2009). Conducting the train of thought: working memory capacity, goal neglect, and mind wandering in an executive-control task. *Journal of Experimental Psychology: Learning, Memory, and Cognition*, *35*(1), 196–204.

Mehigan, T. J., Barry, M., Kehoe, A., & Pitt, I. (2011). Using eye tracking technology to identify visual and verbal learners. In 2011 IEEE International Conference on Multimedia and Expo (pp. 1–6). IEEE.

Mehoudar, E., Arizpe, J., Baker, C. I., & Yovel, G. (2014). Faces in the eye of the beholder: Unique and stable eye scanning patterns of individual observers. *Journal of Vision*, *14*(7), 1–11.

Meyberg, S., Werkle-Bergner, M., Sommer, W., & Dimigen, O. (2015). Microsaccade-related brain potentials signal the focus of visuospatial attention. *NeuroImage*, *104*, 79–88.

Meyer, A. S. (2004). The use of eye tracking in studies of sentence generation. In Henderson, J. & Ferreira, F. (Eds.). *The interface of language, vision, and action: Eye movements and the visual world*, (pp. 191–211). Taylor and Francis.

Meyer, A. S., Sleiderink, A. M., & Levelt, W. J. (1998). Viewing and naming objects: Eye movements during noun phrase production. *Cognition*, 66(2), B25–B33.

Meyer, D. E., & Schvaneveldt, R. W. (1971). Facilitation in recognizing pairs of words: evidence of a dependence between retrieval operations. *Journal of Experimental Psychology*, *90*(2), 227–234.

Mills, C., Gregg, J., Bixler, R., & D'Mello, S. K. (2020). Eye-Mind reader: an intelligent reading interface that promotes long-term comprehension by detecting and responding to mind wandering. *Human-Computer Interaction*, *36*(4), 1–27.

Mirault, J., Yeaton, J., Broqua, F., Dufau, S., Holcomb, P. J., & Grainger, J. (2020). Parafoveal-on-foveal repetition effects in sentence reading: A co-registered eye-tracking and electroencephalogram study. *Psychophysiology*, e13553.

Mittner, M., Boekel, W., Tucker, A. M., Turner, B. M., Heathcote, A., & Forstmann, B. U. (2014). When the brain takes a break: a model-based analysis of mind wandering. *Journal of Neuroscience*, *34*(49), 16286–16295.

Moers, C., Meyer, A., & Janse, E. (2017). Effects of word frequency and transitional probability on word reading durations of younger and older speakers. *Language and Speech*, *60*(2), 289–317.

Mooneyham, B. W., & Schooler, J. W. (2013). The costs and benefits of mind-wandering: a review. *Canadian Journal of Experimental Psychology/Revue canadienne de psychologie expérimentale*, *67*(1), 11.

Moore, C., & Dunham, P. J. (1995). *Joint Attention: Its Origin and Role in Development*. Hillsdale: Lawrence Erlbaum.

Montero Perez, M., Peters, E., & Desmet, P. (2015). Enhancing vocabulary learning through captioned Video: An eye-tracking study. *The Modern Language Journal*, *99*(2), 308–328.

Montero Perez, M., Van Den Noortgate, W., & Desmet, P. (2013). Captioned video for L2 listening and vocabulary learning: A meta-analysis. *System*, *41*(3), 720–739.

Morimoto, C. H., & Mimica, M. R. (2005). Eye gaze tracking techniques for interactive applications. *Computer Vision and Image Understanding*, *98*(1), 4–24.

Mowlem, F., Skirrow, C., Reid, P., Maltezos, S., Nijjar, S., Merwood, A., Barker, E., Cooper, R., Kuntsi, J., & Asherson, P. (2016). Validation of the mind excessively wandering scale and the relationship of mind wandering to impairment in adult ADHD. *Journal of Attention Disorders*, *23*(6), 624–634. https://doi.org/10.1177/1087054716651927

Mrazek, M. D., Smallwood, J., & Schooler, J. W. (2012). Mindfulness and mind-wandering: finding convergence through opposing constructs. *Emotion*, *12*(3), 442–448.

Mundy, P., & Newell, L. (2007). Attention, joint attention, and social cognition. *Current Directions in Psychological Science*, *16*(5), 269–274.

Muñoz, C. (2017). The role of age and proficiency in subtitle reading. An eye-tracking study. *System*, *67*, 77–86.

Mutlu, B., Shiwa, T., Kanda, T., Ishiguro, H., & Hagita, N. (2009). Footing in human-robot conversations: how robots might shape participant roles using gaze cues. In Proceedings of the 4th ACM/IEEE International Conference on Human Robot Interaction (pp. 61–68).

Nakamichi, N., Shima, K., Sakai, M., & Matsumoto, K. I. (2006). Detecting low usability web pages using quantitative data of users' behavior. In Proceedings of the 28th International Conference on Software Engineering (pp. 569–576).

Nakano, Y. I., Reinstein, G., Stocky, T., & Cassell, J. (2003). Towards a model of face-to-face grounding. In Proceedings of the 41st Annual Meeting of the Association for Computational Linguistics (pp. 553–561).

Neider, M. B., & Zelinsky, G. J. (2006). Scene context guides eye movements during visual search. *Vision Research*, *46*(5), 614–621.

Nielsen, J. & Pernice, K. (2010), *Eyetracking web usability*. Pearson: Peachpit.

Nistor, N. & Hernández-García, A. (2018). What types of data are used in learning analytics? An overview of six cases. *Computers in Human Behavior*, *89*, 335–338.

Nordfält, J. (2011). Improving the attention-capturing ability of special displays with the combination effect and the design effect. *Journal of Retailing and Consumer Services*, *18*(3), 169–173.

Noton, D. & Stark, L. (1971). Scanpaths in eye movements during pattern perception. *Science*, *171*(3968), 308–311.

Nüssli, M. A. (2011). Dual eye-tracking methods for the study of remote collaborative problem solving. Unpublished PhD Dissertation. Lausanne, CH.

Oben, B., & Brône, G. (2015). What you see is what you do: on the relation-ship between gaze and gesture in multimodal alignment. *Language and Cognition*, *7*(4), 546–562.

Oben, B., & Brône, G. (2016). Explaining interactive alignment: A multimodal and multifactorial account. *Journal of Pragmatics*, *104*, 32–51.

O'Brien, E. J., Shank, D. M., Myers, J. L., & Rayner, K. (1988). Elaborative inferences during reading: Do they occur on-line? *Journal of Experimental Psychology: Learning, Memory, and Cognition*, *14*(3), 410–420.

Oliva, A. (2005). Gist of the scene. In: Itti, L., Rees, G. & Tsotsos, J. K. (Eds.). *Neurobiology of Attention*, (pp. 251–256). San Diego: Elsevier.

Olmsted-Hawala, E., Holland, T., & Quach, V. (2014). Usability testing. In Bergstrom, J. R. & Schall, A. J. (Eds.). *Eye tracking in user experience design*, (pp. 49–80). Elsevier.

Olsen, A. (2012). The Tobii I-VT fixation filter. *Tobii Technology*. 1–21.

Onifer, W., & Swinney, D. A. (1981). Accessing lexical ambiguities during sentence comprehension: Effects of frequency of meaning and contextual bias. *Memory & Cognition*, 9(3), 225–236.

Orquin, J. L., & Holmqvist, K. (2018). Threats to the validity of eye-movement research in psychology. *Behavior Research Methods*, 50(4), 1645–1656.

Otero-Millan, J., Macknik, S. L., Langston, R. E., & Martinez-Conde, S. (2013). An oculomotor continuum from exploration to fixation. *Proceedings of the National Academy of Sciences of the United States of America*, *110*(15), 6175–6180. https://doi.org/10.1073/pnas.1222715110

Otero-Millan, J., Macknik, S. L., & Martinez-Conde, S. (2014). Fixational eye movements and binocular vision. *Frontiers in Integrative Neuroscience*, *8*, 52. https://doi.org/10.3389/fnint.2014.00052

Papadatou-Pastou, M., Touloumakos, A. K., Koutouveli, C., & Barrable, A. (2020). The learning styles neuromyth: When the same term means different things to different teachers. *European Journal of Psychology of Education*, *36*(2), 511–553.

Paré, M., Richler, R. C., ten Hove, M. et al. (2003). Gaze behavior in audiovisual speech perception: The influence of ocular fixations on the McGurk effect. *Perception & Psychophysics*, *65*, 553–567.

Parker, A. J., Slattery, T. J., & Kirkby, J. A. (2019). Return-sweep saccades during reading in adults and children. *Vision Research*, *155*, 35–43.

Pasqualotti, L., & Baccino, T. (2014). Online advertisement: how are visual strategies affected by the distance and the animation of banners?. *Frontiers in Psychology*, *5*, 211.

Paterson, K. B., McGowan, V. A., White, S. J., Malik, S., Abedipour, L., & Jordan, T. R. (2014). Reading direction and the central perceptual span in Urdu and English. *PloS One*, *9*(2), e88358.

Paxton, A., & Dale, R. (2017). Interpersonal movement synchrony responds to high-and low-level conversational constraints. *Frontiers in psychology, 8*(1135), 1–16.

Peißl, S., Wickens, C. D., & Baruah, R. (2018). Eye-tracking measures in aviation: A selective literature review. *The International Journal of Aerospace Psychology*, *28*(3–4), 98–112.

Pelachaud, C. (2015, May). Greta: an interactive expressive embodied conversational agent. In Proceedings of the 2015 International Conference on Autonomous Agents and Multiagent Systems (pp. 5–5).

Perdomo, M., & Kaan, E. (2019). Prosodic cues in second-language speech processing: A visual world eye-tracking study. *Second Language Research*, 0267658319879196, 1–27.

Pfeiffer, T., & Renner, P. (2018). Quantifying the interplay of gaze and gesture in deixis using an experimental-simulative approach. In *Eye-tracking in interaction*. In Brône, G. & Oben, B. (Eds.). *Studies on the role of eye gaze in dialogue*, (pp. 109–138). Benjamins.

Pfeiffer, U. J., Schilbach, L., Jording, M., Timmermans, B., Bente, G., & Vogeley, K. (2012). Eyes on the mind: investigating the influence of gaze dynamics on

the perception of others in real-time social interaction. *Frontiers in Psychology, 3*(537), 1–11. https://doi.org/10.3389/fpsyg.2012.00537
Pfeiffer, U. J., Vogeley, K., & Schilbach, L. (2013). From gaze cueing to dual eye-tracking: novel approaches to investigate the neural correlates of gaze in social interaction. *Neuroscience & Biobehavioral Reviews*, *37*(10), 2516–2528.
Pickering, L. (2018). . *Discourse intonation: A discourse-pragmatic approach to teaching the pronunciation of English*. University of Michigan Press.
Pickering, L., Hu, G., & Baker, A. (2012). The pragmatic function of intonation: Cueing agreement and disagreement in spoken English discourse and implications for ELT. In Romeo, J. (Ed.). *Pragmatics, prosody & ELT*, (pp. 199–218). Springer.
Pickering, M. J., & Gambi, C. (2018). Predicting while comprehending language: A theory and review. *Psychological Bulletin*, *144*(10), 1002–1044.
Pickering, M. J., & Garrod, S. (2004). Toward a mechanistic psychology of dialogue. *Behavioral and brain sciences*, *27*(2), 169–190.
Pieters, R., & Warlop, L. (1999). Visual attention during brand choice: The im-pact of time pressure and task motivation. *International Journal of Research in Marketing*, *16*(1), 1–16.
Pieters, R., & Wedel, M. (2004). Attention capture and transfer in advertising: Brand, pictorial, and text-size effects. *Journal of Marketing*, *68*(2), 36–50.
Pieters, R., & Wedel, M. (2012). Ad gist: Ad communication in a single eye fixation. *Marketing Science*, *31*(1), 59–73.
Pieters, R., Wedel, M., & Zhang, J. (2007). Optimal feature advertising design under competitive clutter. *Management Science*, 53(11), 1815–1828.
Pluzyczka, M. (2018). The first hundred years: a history of eye tracking as a research method. *Applied Linguistics Papers*, *4*(25), 101–116.
Poggi, I., Pelachaud, C., & De Rosis, F. (2000). Eye communication in a conversational 3D synthetic agent. *AI Communications*, *13*(3), 169–181.
Poggi, I., & Pelachaud, C. (2002). Signals and meanings of gaze in animated faces. *Advances in Consciousness Research*, *35*, 133–144.
Poggi, I., Pelachaud, C., de Rosis, F., Carofiglio, V., & De Carolis, B. (2005). Greta. a believable embodied conversational agent. In Stock, O. & Zancanaro, M. (Eds.). *Multimodal intelligent information presentation*, (pp. 3–25). Dordrecht: Springer.
Pollatsek, A., Lesch, M., Morris, R. K., & Rayner, K. (1992). Phonological codes are used in integrating information across saccades in word identification and reading. *Journal of Experimental Psychology: Human Perception and Performance*, *18*(1), 148–162.
Pollatsek, A., Reichle, E. D., & Rayner, K. (2003). Modeling eye movements in reading: Extensions of the EZ Reader model. In Hyönä, J., Radach, R., & Deubel, H. (Eds.). *The Mind's Eye: Cognitive and Applied Aspect of Eye Movement Research*, (pp. 361–390). North-Holland.
Pönkänen, L. M., & Hietanen, J. K. (2012). Eye contact with neutral and smiling faces: effects on autonomic responses and frontal EEG asymmetry. *Frontiers in Human Neuroscience*, 6(122), 1–9.
Popper, K. R. (1959). *The Logic of Scientific Discovery*. New York: Hutchinson.
Posner, M. I., Snyder, C. R., & Davidson, B. J. (1980). Attention and the detection of signals. *Journal of experimental psychology: General*, *109*(2), 160–174.

Pozzan, L., & Trueswell, J. C. (2016). Second language processing and revision of garden-path sentences: a visual word study. *Bilingualism*, *19*(3), 636–643.

Price, K. (1983). Effects of captioned television on adult ESL learners. *MATSOL Newsletter*, *12*(2), 1,8.

Qin, Z., Tremblay, A., & Zhang, J. (2019). Influence of within-category tonal information in the recognition of Mandarin-Chinese words by native and non-native listeners: An eye-tracking study. *Journal of Phonetics*, *73*, 144–157.

Qvarfordt, P., Beymer, D., & Zhai, S. (2005). Realtourist—a study of augmenting human-human and human-computer dialogue with eye-gaze overlay. In IFIP Conference on Human-Computer Interaction (pp. 767–780). Berlin, Heidelberg: Springer.

Rajakumari, B., & Selvi, N. S. (2016). HCI and eye tracking: Emotion recognition using hidden markov model. *International Journal of Computer Science and Network Security (IJCSNS)*, *16*(3), 74–77.

Randall, J. G., Oswald, F. L., & Beier, M. E. (2014). Mind-wandering, cognition, and performance: A theory-driven meta-analysis of attention regulation. *Psychological bulletin*, *140*(6), 1411–1431.

Rao, R. P., Zelinsky, G. J., Hayhoe, M. M., & Ballard, D. H. (2002). Eye movements in iconic visual search. *Vision Research*, 42(11), 1447–1463.

Ratiu, I., & Azuma, T. (2015). Working memory capacity: Is there a bilingual advantage?. *Journal of Cognitive Psychology*, *27*(1), 1–11.

Raudonis, V., Dervinis, G., Vilkauskas, A., Paulauskaite, A., & Kersulyte, G. (2013). Evaluation of human emotion from eye motions. *International Journal of Advanced Computer Science and Applications*, *4*(8). 79–84.

Rayner, K. (1998). Eye movements in reading and information processing: 20 years of research. *Psychological bulletin*, *124*(3), 372.

Rayner, K. (2009a). Eye movements and attention in reading, scene perception, and visual search. *The Quarterly Journal of Experimental Psychology*, *62*(8), 1457–1506. https://doi.org/10.1080/17470210902816461

Rayner, K. (2009b). Eye movements in reading: Models and data. *Journal of Eye Movement Research, 2*(5), 1.

Rayner, K. (2014). The gaze-contingent moving window in reading: Develop-ment and review, *Visual Cognition*, *22*(3–4), 242–258.

Rayner, K., & Castelhano, M. S. (2008). Eye movements during reading, scene perception, visual search, and while looking at print advertisements. In Wedel, M. & Pieters, R. (Eds.). *Visual marketing: From attention to action* (p. 9–42). Taylor & Francis/Lawrence Erlbaum.

Rayner, K., & Duffy, S. A. (1986). Lexical complexity and fixation times in reading: Effects of word frequency, verb complexity, and lexical ambiguity. *Memory & Cognition*, *14*(3), 191–201.

Rayner, K., Inhoff, A. W., Morrison, R. E., Slowiaczek, M. L., & Bertera, J. H. (1981). Masking of foveal and parafoveal vision during eye fixations in reading. *Journal of Experimental Psychology: Human Perception and Performance*, *7*(1), 167–179.

Rayner, K., Kambe, G., & Duffy, S. A. (2000). The effect of clause wrap-up on eye movements during reading. *The Quarterly Journal of Experimental Psychology: Section A*, *53*(4), 1061–1080.

Rayner, K & Pollatsek, A. (2006). Eye-movement control in reading. In Traxler, M. J. & Gernsbacher, M. A. (Eds.). *Handbook of psycholinguistics*, (pp. 613–657). Elsevier.

Rayner, K., & Reichle, E. D. (2010). Models of the reading process. *Wiley Interdisciplinary Reviews: Cognitive Science*, *1*(6), 787–799.

Rayner, K., Slattery, T. J., & Bélanger, N. N. (2010). Eye movements, the perceptual span, and reading speed. *Psychonomic Bulletin & Review*, *17*(6), 834–839.

Rayner, K., Warren, T., Juhasz, B. J., & Liversedge, S. P. (2004). The effect of plausibility on eye movements in reading. *Journal of Experimental Psychology: Learning, Memory, and Cognition*, *30*(6), 1290–1301.

Records, R. E. (Ed.) (1979). *Physiology of the Human Eye and Visual System*. Cambridge: Harper& Row.

Reed, J. B. & Meyer, R. J. (2007). Edmund Burke Huey (1870–1913): A brief life with an enduring legacy. In Israel, S. E., & Monaghan, E. J. (Eds.). *Shaping the reading field: The impact of early reading pioneers, scientific research, and progressive ideas*. (pp. 159–175). Newark: International Reading Association.

Reichle, E. D. (2021). *Computational models of reading. A handbook*. Oxford University Press.

Reichle, E. D., & Laurent, P. A. (2006). Using reinforcement learning to understand the emergence of "intelligent" eye-movement behavior during reading. *Psychological Review*, *113*(2), 390–408.

Reichle, E. D., Pollatsek, A., Fisher, D. L., & Rayner, K. (1998). Toward a model of eye movement control in reading. *Psychological Review*, *105*(1), 125–157.

Reichle, E. D., Rayner, K., & Pollatsek, A. (2003). The EZ Reader model of eye-movement control in reading: Comparisons to other models. *Behavioral and Brain Sciences*, *26*(4), 445–476.

Reichle, E. D., Reineberg, A. E., & Schooler, J. W. (2010). Eye movements during mindless reading. *Psychological Science*, *21*(9), 1300–1310.

Reinagel, P., & Zador, A. M. (1999). Natural scene statistics at the centre of gaze. *Network: Computation in Neural Systems*, *10*(4), 341–350.

Remington, R. W., Burt, J. S., & Becker, S. I. (2018). The curious case of spillover: Does it tell us much about saccade timing in reading?. *Attention, Perception, & Psychophysics*, *80*(7), 1683–1690.

Richardson, D. C., & Dale, R. (2005). Looking to understand: The coupling between speakers' and listeners' eye movements and its relationship to discourse comprehension. *Cognitive Science*, *29*(6), 1045–1060.

Richardson, D. C., & Spivey, M. J. (2004). Eye tracking: Characteristics and methods. In Wnek, G. E., & Bowlin, G. L. (Eds.). *Encyclopedia of biomaterials and biomedical engineering*, 3, 1028–1042.

Richter, J., Scheiter, K., & Eitel, A. (2016). Signaling text-picture relations in multimedia learning: A comprehensive meta-analysis. *Educational Research Review*, *17*, 19–36. https://doi.org/10.1016/j.edurev.2015.12.003

Riener, C. & Willingham, D. (2010). The myth of learning styles. *Change: The Magazine of Higher Learning*, *42*(5). 32–35.

Risko, E. F., Anderson, N., Sarwal, A., Engelhardt, M., & Kingstone, A. (2012). Everyday attention: Variation in mind wandering and memory in a lecture. *Applied Cognitive Psychology*, *26*(2), 234–242.

Risse, S., Hohenstein, S., Kliegl, R., & Engbert, R. (2014). A theoretical analysis of the perceptual span based on SWIFT simulations of the n+ 2 boundary paradigm. *Visual Cognition*, *22*(3–4), 283–308.

Ritschel, H., Aslan, I., Sedlbauer, D., & André, E. (2019). Irony man: Augmenting a social robot with the ability to use irony in multimodal communication with humans. In Proceedings of the 18th International Conference on Autonomous Agents and MultiAgent Systems (pp. 86–94).

Roberts, D., Shelhamer, M., & Wong, A. (2008). A new "wireless" search-coil system. In Proceedings of the 2008 Symposium on Eye Tracking Research & Applications (pp. 197–204).

Roberts, L., & Siyanova-Chanturia, A. (2013). Using eye-tracking to investigate topics in L2 acquisition and L2 processing. *Studies in Second Language Acquisition*, *35*(2), 213–235.

Robinson, D. A. (1963). A method of measuring eye movement using a scleral search coil in a magnetic field. *IEEE Transactions on bio-medical electronics*, *10*(4), 137–145.

Roda, C., & Thomas, J. (2006). Attention aware systems: Theories, applications, and research agenda. *Computers in Human Behavior*, *22*(4), 557–587.

Rogers, S. L., Guidetti, O., Speelman, C. P., Longmuir, M., & Phillips, R. (2019). Contact is in the eye of the beholder: The eye contact illusion. *Perception*, *48*(3), 248–252.

Rogers, S. L., Speelman, C. P., Guidetti, O., & Longmuir, M. (2018). Using dual eye tracking to uncover personal gaze patterns during social interaction. *Nature. Scientific reports*, *8*(1), 1–9.

Rohrer, D., & Pashler, H. (2012). Learning styles: Where's the evidence? *Medical Research*, *46*(7), 634–635.

Rosengrant, D., Hearrington, D., Alvarado, K., & Keeble, D. (2011). Following student gaze patterns in physical science lectures. In Physics Education Research Conference Proceedings, 1413, (pp. 323–326). American Institute of Physics.

Rosengrant, D., Hearrington, D., & O'Brien, J. (2020). Investigating student sustained attention in a guided inquiry lecture course using an eye tracker. *Educational Psychology Review*. https://doi.org/10.1007/s10648-020-09540-2

Rosner, A., Franke, T., Platten, F. & Attig, C. (2019). Eye movements in vehicle control. In Klein, C. & Ettinger, U. (Eds.). *Eye movement research*. Springer.

Rossano, F. (2013). Gaze in conversation. In Sidnell, J., & Stivers, T. (Eds.). *The handbook of conversation analysis*, (pp. 308–329). Wiley.

Rothensee, M., & Reiter, P. (2019). Neuromarketing. In Klein, C. & Ettinger, U. (Eds.). *Eye Movement Research*, (pp. 819–855). Cham: Springer.

Ruhland, K., Peters, C. E., Andrist, S., Badler, J. B., Badler, N. I., Gleicher, M., B. Mutlu, & McDonnell, R. (2015). A review of eye gaze in virtual agents, social robotics and HCI: Behaviour generation, user interaction and perception. *Computer Graphics Forum*, *34*(6). 299–326.

Sacks, H., Schegloff, E., & Jefferson, G., (1974). A simplest systematics for the organization of turn-taking for conversation. *Language*, *50*(4), 696–735.

Saito, D. N., Tanabe, H. C., Izuma, K., Hayashi, M. J., Morito, Y., Komeda, H., . . . Sadato, N. (2010). ?Stay tuned?: inter-individual neural synchronization during mutual gaze and joint attention. *Frontiers in Integrative Neuroscience,* 4(127), 1–12.

Salverda, A. P., & Tanenhaus, M. K. (2010). Tracking the time course of orthographic information in spoken-word recognition. *Journal of Experimental Psychology: Learning, Memory, and Cognition*, *36*(5), 1108–1117.

Salvucci, D. D., & Goldberg, J. H. (2000). Identifying fixations and saccades in eye-tracking protocols. In Proceedings of the 2000 Symposium on Eye Tracking Research & Applications (pp. 71–78).

Sangin, M., Molinari, G., Nüssli, M. A., & Dillenbourg, P. (2008). How learners use awareness cues about their peer's knowledge? Insights from synchronized eye-tracking data. In Kanselaar, G., Jonker, V., Kirschner, P. A., Prins, F. J. (Eds.). *International perspectives in the learning sciences: Cre8ing a learning world.* Proceedings of the Eighth International Conference for the Learning Sciences – ICLS 2008, 2 (pp. 287–294). International Society of the Learning Sciences.

Santesteban, M., & Schwieter, J. W. (2020). Lexical Selection and Competition in Bilinguals. In Heredia, R. R., & Cie'licka, A. B. (Eds.). *Bilingual lexical ambiguity resolution*, (pp. 126–156). Cambridge.

Schackwitz, A. (1913). Apparat zur Aufzeichnung der Augenbewegungen beim zusammenhängenden Lesen. *Zeitchriften für Psychologie*, *63*, 442–453.

Schäfer, A., Isomura, T., Reis, G., Watanabe, K., & Stricker, D. (2020). MutualEyeContact: A conversation analysis tool with focus on eye contact. In ACM Symposium on Eye Tracking Research and Applications (pp. 1–5). Association for Computing Machinery.

Scheiter, K., & Eitel, A. (2015). Signals foster multimedia learning by supporting integration of highlighted text and diagram elements. *Learning and Instruction*, *36*, 11–26.

Schilbach, L., Wilms, M., Eickhoff, S. B., Romanzetti, S., Tepest, R., Bente, G., . . . & Vogeley, K. (2010). Minds made for sharing: initiating joint attention recruits reward-related neurocircuitry. *Journal of cognitive neuroscience*, *22*(12), 2702–2715.

Schiller, P. H., & Tehovnik, E. J. (2015). *Vision and the visual system.* Oxford University Press.

Schmidt, R. (1990). The role of consciousness in second language learning. *Applied Linguistics*, *11*, 129–158.

Schmidt, R. (2012). Attention, awareness, and individual differences in language learning. In Chan, W. M., Chin, K. N., Bhatt, S., & Walker, I. (Eds.). *Perspectives on individual characteristics and foreign language education*, (pp. 27–50). DeGruyter.

Schneider, B., Abu-El-Haija, S., Reesman, J., & Pea, R. (2013). Toward collaboration sensing: Applying network analysis techniques to collaborative eye-tracking data. In Proceedings of the Third International Conference on Learning Analytics and Knowledge (pp. 107–111).

Schotter, E. R., Angele, B., & Rayner, K. (2012). Parafoveal processing in reading. *Attention, Perception, & Psychophysics*, *74*(1), 5–35.

Sedivy, J. C., Tanenhaus, M. K., Chambers, C. G., & Carlson, G. N. (1999). Achieving incremental semantic interpretation through contextual representation. *Cognition*, *71*(2), 109–147.

Sedivy, J. C., Tanenhaus, M. K., Spivey-Knowlton, M., Eberhard, K., & Carlson, G. (1995). Using intonationally-marked presuppositional information in on-line language processing: Evidence from eye movements to a visual model. In Moore, J. D. & Lehman, J. F. (Eds.). *Proceedings of the 17th Annual Conference of the Cognitive Science Society*, (pp. 375–380). Erlbaum Hillsdale, NJ.

Seli, P., Risko, E. F., Purdon, C., & Smilek, D. (2017). Intrusive thoughts: Linking spontaneous mind wandering and OCD symptomatology. *Psychological research*, *81*(2), 392–398.

Seli, P., Risko, E. F., Smilek, D., & Schacter, D. L. (2016). Mind-wandering with and without intention. *Trends in Cognitive Sciences*, 20(8), 605–617.

Seli, P., Schacter, D. L., Risko, E. F., & Smilek, D. (2019). Increasing participant motivation reduces rates of intentional and unintentional mind wandering. *Psychological Research*, *83*(5), 1057–1069.

Seli, P., Smallwood, J., Cheyne, J. A., & Smilek, D. (2015). On the relation of mind wandering and ADHD symptomatology. *Psychonomic Bulletin & Review*, *22*(3), 629–636.

Senju, A., & Johnson, M. H. (2009). The eye contact effect: Mechanisms and development. *Trends in Cognitive Sciences*, *13*(3), 127–134.

Shackel, B. (1960). Note on mobile eye viewpoint recording. *Journal of the Optical Society of America*, *50*(8), 763–768.

Sharafi, Z., Soh, Z., & Guéhéneuc, Y. G. (2015). A systematic literature review on the usage of eye-tracking in software engineering. *Information and Software Technology*, *67*, 79–107.

Sharif, B., & Maletic, J. I. (2010). An eye tracking study on camelCase and under_score identifier styles. In 2010 IEEE 18th International Conference on Program Comprehension (pp. 196–205). IEEE.

Sharwood Smith, M. (1993). Input enhancement in instructed SLA: Theoretical bases. *Studies in Second Language Acquisition*, *15*(2), 165–179.

Sheldon, K. M., Prentice, M., & Halusic, M. (2015). The experiential incompatibility of mindfulness and flow absorption. *Social Psychological and Personality Science*, *6*(3), 276–283.

Shepherd, S. V. (2010). Following gaze: Gaze-following behavior as a window into social cognition. *Frontiers in Integrative Neuroscience*, *4*, 1–13.

Shimojo, S., Simion, C., Shimojo, E., & Scheier, C. (2003). Gaze bias both reflects and influences preference. *Nature neuroscience*, *6*(12), 1317–1322.

Shook, A., & Marian, V. (2019). Covert co-activation of bilinguals' non-target language: Phonological competition from translations. *Linguistic Approaches to Bilingualism*, *9*(2), 228–252.

Sibert, J. L., Gokturk, M., & Lavine, R. A. (2000). The reading assistant: eye gaze triggered auditory prompting for reading remediation. In Proceedings of the 13th Annual ACM Symposium on User Interface Software and Technology (pp. 101–107).

Sidnell, J., & Stivers, T. (Eds.). (2012). *The handbook of conversation analysis*. Wiley.

Simarro Vázquez, M., El Khatib, N., Hamrick, P., & Attardo, S. (2020). On the order of processing of humorous tweets with visual and verbal elements. *Internet Pragmatics*. https://doi.org/10.1075/ip.00060.sim. 1–26.

Simons, D. & Chabris, C. F. (1999). Gorillas in our midst: Sustained inattentional blindness for dynamic events. *Perception*, *28*(9), 1059–1074.

Simonsen, H. K. (2011). User consultation behaviour in internet dictionaries: An eye-tracking study. *HERMES-Journal of Language and Communication in Business*, *46*, 75–101.

Sims, S., & Conati, C. (2020). A neural architecture for detecting confusion in eye-tracking data. *arXiv preprint* arXiv:2003.06434.

Singh, N., & Mishra, R. K. (2013). Second language proficiency modulates conflict-monitoring in an oculomotor Stroop task: Evidence from Hindi-English bilinguals. *Frontiers in psychology*, *4*(322), 1–10.

Sisk, C. A., Remington, R. W., & Jiang, Y. V. (2019). Mechanisms of contextual cueing: A tutorial review. *Attention, Perception, & Psychophysics*, *81*(8), 2571–2589.

Skantze, G., Hjalmarsson, A., & Oertel, C. (2014). Turn-taking, feedback and joint attention in situated human-robot interaction. *Speech Communication*, *65*, 50–66.

Smallwood, J. (2011). Mind-wandering while reading: Attentional decoupling, mindless reading and the cascade model of inattention. *Language and Linguistics Compass*, *5*(2), 63–77.

Smallwood, J. (2015). Mind wandering and attention. In Fawcett, J. M., Risko, E. F., & Kingstone, A. (Eds.). *The handbook of attention* (pp. 233–255). MIT Press.

Smallwood, J., Brown, K. S., Baird, B., Mrazek, M. D., Franklin, M. S., & Schooler, J. W. (2012). Insulation for daydreams: a role for tonic norepinephrine in the facilitation of internally guided thought. *PloS One*, 7(4), e33706.

Smallwood, J., Fishman, D. J., & Schooler, J. W. (2007). Counting the cost of an absent mind: Mind wandering as an underrecognized influence on educational performance. *Psychonomic Bulletin & Review*, *14*(2), 230–236.

Smallwood, J., McSpadden, M., & Schooler, J. W. (2008). When attention matters: The curious incident of the wandering mind. *Memory & Cognition*, *36*(6), 1144–1150.

Smallwood, J. & Schooler, J. W. (2015). The science of mind wandering: Empirically navigating the stream of consciousness. *Annual Review of Psychology*, *66*(1). 487–518.

Smilek, D., Carriere, J. S., & Cheyne, J. A. (2010). Out of mind, out of sight: eye blinking as indicator and embodiment of mind wandering. *Psychological Science*, *21*(6), 786–789.

Snowden, R. J., Thompson, P., & Troscianko, T. (2012). *Basic vision: An introduction to visual perception* (2nd ed.). Oxford University Press.

Specker, E. (2008). L1/L2 eye movement reading of closed captioning: A multimodal analysis of multimodal use. Unpublished doctoral dissertation. University of Arizona.

Spielman, R. M., Dumper, K., Jenkins, W., Lacombe, A., Lovett, M., & Perlmutter, M. (2019). *Psychology*. https://opentextbc.ca/psychologyopenstax/

Spivey, M. J., & Huette, S. (2016). Toward a situated view of language. In Knoeferle, P., Pyykkönen-Klauck, P., & Crocker, M. W. (Eds.). *Visually situated language comprehension*, (pp. 1–30). Benjamins.

Spivey, M. J., & Tanenhaus, M. K. (1998). Syntactic ambiguity resolution in discourse: Modeling the effects of referential context and lexical frequency. *Journal of Experimental Psychology: Learning, Memory, and Cognition*, *24*(6), 1521–1543.

Spotorno, N., Cheylus, A., Van Der Henst, J. B., & Noveck, I. A. (2013). What's behind a P600? Integration operations during irony processing. *PloS One*, *8*(6), e66839.

Stanley, G., & Martin, D. S. (1968). Eye-contact and the recall of material involving competitive and noncompetitive associations. *Psychonomic Science*, *13*(6), 337–338.

Staub, A. (2015). The effect of lexical predictability on eye movements in reading: Critical review and theoretical interpretation. *Language and Linguistics Compass*, *9*(8), 311–327.

Staudte, M., & Crocker, M. W. (2011). Investigating joint attention mechanisms through spoken human–robot interaction. *Cognition*, *120*(2), 268–291.

Staudte, M., & Crocker, M. W. (2018). On the role of gaze for successful and efficient communication. In Brône, G., & Oben, B. (Eds.). *Eye-tracking in interaction: Studies on the role of eye gaze in dialogue*. (pp. 91–106).Benjamins.

Stenfors, I., Morén, J., & Balkenius, C. (2003). Behavioural strategies in web interaction: A view from eye-movement research. In Hyönä, J., Radach, R., Deubel, H. (Eds.). *The mind's eye: Cognitive and applied aspects of eye movement research*, (pp. 633–644). Elsevier.

Stifter, C. A., & Moyer, D. (1991). The regulation of positive affect: Gaze aversion activity during mother-infant interaction. *Infant Behavior and Development*, *14*(1), 111–123.

Stowe, L. A., Kaan, E., Sabourin, L., & Taylor, R. C. (2018). The sentence wrap-up dogma. *Cognition*, *176*, 232–247.

Stratton, G. M. (1902). Studies from the psychological laboratory of the University of California. *Psychological Review, 9*(5), 433–443.

Stratton, G. M. (1906). Symmetry, linear illusions, and the movements of the eye. *Psychological Review*, *13*(2), 82–96.

Strauch, C., Greiter, L., & Huckauf, A. (2018). Pupil dilation but not microsaccade rate robustly reveals decision formation. *Scientific Reports,* 8(1), 1–9.

Strayer, D. L. (2015). Is the technology in your car driving you to distraction?. *Policy Insights from the Behavioral and Brain Sciences*, 2(1), 157–165.

Stroop, J. R. (1935). Studies of interference in serial verbal reactions. *Journal of Experimental Psychology*, *18*(6), 643–662.

Styles, E. A. (2006). *The psychology of attention* (2nd ed.). Psychology Press.

Sullivan, G. M., & Feinn, R. (2012). Using effect size-or why the P value is not enough. *Journal of Graduate Medical Education*, *4*(3), 279–282. https://doi.org/10.4300/JGME-D-12-00156.1

Sun, C., & Breheny, R. (2019). Another look at the online processing of scalar inferences: An investigation of conflicting findings from visual-world eye-tracking studies. *Language, Cognition and Neuroscience*, *35*(8), 949–979.

Szczepek Reed, B. (2006). *Prosodic orientation in English conversation*. Palgrave.

Szpunar, K. K., Khan, N. Y., & Schacter, D. L. (2013). Interpolated memory tests reduce mind wandering and improve learning of online lectures. *Proceedings of the National Academy of Sciences*, *110*(16), 6313–6317.

Szpunar, K. K., Jing, H. G., & Schacter, D. L. (2014). Overcoming overconfidence in learning from video-recorded lectures: Implications of interpolated testing for online education. *Journal of Applied Research in Memory and Cognition*, *3*, 161–164.

Tanenhaus, M. K., Spivey-Knowlton, M. J., Eberhard, K. M., & Sedivy, J. C. (1995). Integration of visual and linguistic information in spoken language comprehension. *Science*, *268*(5217), 1632–1634.

Tatler, B. W., Wade, N. J., Kwan, H., Findlay, J. M., & Velichkovsky, B. M. (2010). Yarbus, eye movements, and vision. *i-Perception*, *1*(1), 7–27.

Tavano, E., & Kaiser, E. (2009). The cost of being cooperative: Evidence of effort in the processing of scalar implicature. In Proceedings from the Annual Meeting of the Chicago Linguistic Society (Vol. 45, pp. 593–607). Chicago Linguistic Society.

Taylor, W. L. (1953). "Cloze procedure": A new tool for measuring readability. *Journalism quarterly*, *30*(4), 415–433.

Theeuwes, J., & Burger, R. (1998). Attentional control during visual search: the effect of irrelevant singletons. *Journal of Experimental Psychology: Human Perception and Performance*, *24*(5), 1342–1353.

Thórisson, K. R. (2002). Natural turn-taking needs no manual: Computational theory and model, from perception to action. In Granström, B., House, D., Karlsson, I. (Eds.). *Multimodality in language and speech systems* (pp. 173–207). Dordrecht: Springer.

Thorpe, S., Fize, D., & Marlot, C. (1996). Speed of processing in the human visual system. *Nature*, *381*(6582), 520–522.

Tipples, J., & Pecchinenda, A. (2019). A closer look at the size of the gaze-liking effect: a preregistered replication. *Cognition and Emotion*, *33*(3), 623–629.

Titz, J., Scholz, A., & Sedlmeier, P. (2018). Comparing eye trackers by correlating their eye-metric data. *Behavior Research Methods*, *50*(5), 1853–1863.

Tono, Y. (2011). Application of eye-tracking in EFL learners' dictionary look-up process research. *International Journal of Lexicography*, *24*(1), 124–153.

Töpel, A. (2014). Review of research into the use of electronic dictionaries. In Müller-Spitzer, C. (Ed.). *Using online dictionaries*, (pp. 13–54). De Gruyter.

Torralba, A., Oliva, A., Castelhano, M. S., & Henderson, J. M. (2006). Contextual guidance of eye movements and attention in real-world scenes: the role of global features in object search. *Psychological review*, *113*(4), 766–786.

Troxler, I. P. V. (1804). Ueber das Verschwinden gegebener Gegenstande inner-halb unseres Gesichtskreises. *Ophthalmologische bibliothek*, *2*, 1–53.

Tsianos, N., Germanakos, P., Lekkas, Z., Mourlas, C., & Samaras, G. (2009). Eye-tracking users' behavior in relation to cognitive style within an e-learning environment. In 2009 Ninth IEEE International Conference on Advanced Learning Technologies (pp. 329–333). IEEE.

Tzourio-Mazoyer, N. & Seghier, M. L. (2016). The neural bases of hemispheric specialization. *Neuropsychologia*, *96*, 319–324.

Unsworth, N., & Robison, M. K. (2018). Tracking arousal state and mind wandering with pupillometry. *Cognitive, Affective, & Behavioral Neuroscience*, *18*(4), 638–664.

Uzzaman, S., & Joordens, S. (2011). The eyes know what you are thinking: eye movements as an objective measure of mind wandering. *Consciousness and Cognition*, *20*(4), 1882–1886.

van den Bogert, N., van Bruggen, J., Kostons, D., & Jochems, W. (2014). First steps into understanding teachers' visual perception of classroom events. *Teaching and Teacher Education*, *37*, 208–216.

Vanderplank, R. (2016). 'Effects of' and 'effects with' captions: How exactly does watching a TV programme with same-language subtitles make a difference to language learners?. *Language Teaching*, *49*(2), 235–250.

van der Wel, P., & van Steenbergen, H. (2018). Pupil dilation as an index of effort in cognitive control tasks: A review. *Psychonomic Bulletin & Review*, *25*(6), 2005–2015.

Vernon, M. D. (1928). Methods of recording eye movements. *The British Journal of Ophthalmology*, *12*(3), 113–130.

Vrij, A., & Fisher, R. P. (2020). Unravelling the misconception about deception and nervous behavior. *Frontiers in Psychology*, *11*(1377), 1–8.

Wade, N. J. (2010). Pioneers of eye movement research. *i-Perception*, *1*(2), 33–68.

Wade, N. J. (2015). How were eye movements recorded before Yarbus?. *Perception*, *44*(8–9), 851–883.

Wade, N. J., & Tatler, B. W. (2005). *The Moving Tablet of the Eye: The Origins of Modern Eye Movement Research*. Oxford: Oxford University Press.

Wade, N. J., & Tatler, B. W. (2009). Did Javal measure eye movements during reading? *Journal of Eye Movement Research*, 2(5), 5, 1–7.

Wade, N. J., Tatler, B. W., Heller, D. (2003) Dodgeing the issue: Dodge, Javal, Hering, and the measurement of saccades in eye-movement research. *Perception*, *32*, 793–804.

Walcher, S., Körner, C., & Benedek, M. (2017). Looking for ideas: Eye behavior during goal-directed internally focused cognition. *Consciousness and Cognition*, *53*, 165–175.

Wammes, J. D., Seli, P., Cheyne, J. A., Boucher, P. O., & Smilek, D. (2016). Mind wandering during lectures II: Relation to academic performance. *Scholarship of Teaching and Learning in Psychology*, 2(1), 33–48.

Wammes, J. D., Seli, P., & Smilek, D. (2018). Mind-wandering in educational settings. In Fox, K. C. R. & Christoff , K. (Eds.). *The Oxford Handbook of Spontaneous Thought: Mind Wandering, Creativity, and Dreaming*, (pp. 259–271). New York: Oxford University Press.

Wammes, J. D., & Smilek, D. (2017). Examining the influence of lecture format on degree of mind wandering. *Journal of Applied Research in Memory and Cognition*, 6(2), 174–184.

Wang, H., Chignell, M., & Ishizuka, M. (2006). Empathic tutoring software agents using real-time eye tracking. In Proceedings of the 2006 Symposium on Eye Tracking Research & Applications (pp. 73–78).

Warren, T., White, S. J., & Reichle, E. D. (2009). Investigating the causes of wrap-up effects: Evidence from eye movements and E-Z Reader. *Cognition*, *111*(1), 132–137.

Wartenberg, C., & Holmqvist, K. (2005). Daily newspaper layout – designers' predictions of readers' visual behaviour – a case study. *Lund University Cognitive Studies, 126*, 1101–8453.

Weber, A., Grice, M., & Crocker, M. W. (2006). The role of prosody in the interpretation of structural ambiguities: A study of anticipatory eye movements. *Cognition*, *99*(2), B63–B72.[*sic*].

Wedel, M. (2015) Attention research in marketing: A review of eye-tracking studies. In Fawcett, J., Risko, E. F. & Kingstone, A. (Eds.). *The* handbook of attention. MIT Press.

Wedel, M., & Pieters, R. (2000). Eye fixations on advertisements and memory for brands: A model and findings. *Marketing Science*, *19*(4), 297–312.

Wedel, M., & Pieters, R. (2008). *Eye tracking for visual marketing*. Now Publishers Inc.

White, B. J. & Munoz, D. P. (2011) The superior colliculus. In Liversedge, S., Gilchrist, I., & Everling, S. (Eds.). (2011). *The Oxford handbook of eye movements*. (pp. 195–213). Oxford University Press.

Whitford, V., & Titone, D. (2012). Second-language experience modulates first-and second-language word frequency effects: Evidence from eye movement measures of natural paragraph reading. *Psychonomic Bulletin & Review*, *19*(1), 73–80.

Wiggins, B. J., & Christopherson, C. D. (2019). The replication crisis in psychology: An overview for theoretical and philosophical psychology. *Journal of Theoretical and Philosophical Psychology*, *39*(4), 202–217.

Willemse, C., & Wykowska, A. (2019). In natural interaction with embodied robots, we prefer it when they follow our gaze: a gaze-contingent mobile eyetracking study. *Philosophical Transactions of the Royal Society B*, *374*(1771), 20180036.

Williams, J. A., Burns, E. L. & Harmon, E. A., (2009). Insincere utterances and gaze: Eye contact during sarcastic statements. *Perceptual and Motor Skills*, *108*(2), 565–572.

Williams, R., & Morris, R. (2004). Eye movements, word familiarity, and vocabulary acquisition. *European Journal of Cognitive Psychology*, *16*(1–2), 312–339.

Wilms, M., Schilbach, L., Pfeiffer, U., Bente, G., Fink, G. R., & Vogeley, K. (2010).It's in your eyes–using gaze-contingent stimuli to create truly inter-active paradigms for social cognitive and affective neuroscience. *Social Cognitive and Affective Neuroscience*, *5*(1), 98–107.

Winke, P. M. (2013). The effects of input enhancement on grammar learning and comprehension: A modified replication of Lee (2007) with eye-movement data. *Studies in Second Language Acquisition*, *35*(2), 323–352.

Winke, P. M., Godfroid, A., & Gass, S. M. (2013). Introduction to the special issue: Eye-movement recordings in second language research. *Studies in Second Language Acquisition*, *35*(2), 205–212.

Winter, B. (2020). *Statistics for linguists: An introduction using R*. Routledge.

Wolff, C. E., Jarodzka, H., van den Bogert, N., & Boshuizen, H. P. (2016). Teacher vision: expert and novice teachers' perception of problematic classroom management scenes. *Instructional Science*, *44*(3), 243–265.

Wooffitt, R. (2005) *Conversation analysis and discourse analysis*. Sage.

Wright, D. B., & Wareham, G. (2005). Mixing sound and vision: The interaction of auditory and visual information for earwitnesses of a crime scene. *Legal and Criminological Psychology*, *10*(1), 103–108.

Wu, C. C., & Kowler, E. (2013). Timing of saccadic eye movements during visual search for multiple targets. *Journal of Vision*, *13*(11), 1–21.

Wu, E. X. W., Laeng, B., & Magnussen, S. (2012). Through the eyes of the own-race bias: Eye-tracking and pupillometry during face recognition. *Social Neuroscience*, *7*(2), 202–216.

Yacci, M., & Rozanski, E. P. (2012). Student information consumption strategies: implications of the Google effect. In Proceedings of the 2012 iConference (pp. 248–253).

Yarbus, A. L. (1967). *Eye movement and vision*. B. Haigh (Trans.). New York: Plenum. Original Russian edition, 1965, Moscow, Nauka Press.

Yiu, Y. H., Aboulatta, M., Raiser, T., Ophey, L., Flanagin, V. L., zu Eulenburg, P., & Ahmadi, S. A. (2019). DeepVOG: Open-source pupil segmentation and gaze estimation in neuroscience using deep learning. *Journal of Neuroscience Methods*, *324*, 108307.

Young, L. R., & Sheena, D. (1975). Survey of eye movement recording methods. *Behavior Research Methods & Instrumentation*, *7*(5), 397–429.

Zammitto, V., & Steiner, K. (2014). Gaming. In Bergstrom, J. R. & Schall, A. J. (Eds.). *Eye tracking in user experience design*, (pp. 291–310). Elsevier.

Zang, C., Liversedge, S. P., Bai, X., & Yan, G. (2011). Eye movements during Chinese reading. In Liversedge, S., Gilchrist, I., & Everling, S. (Eds.). *The Oxford handbook of eye movements*, (pp. 961–978). Oxford University Press.

Zelinsky, G. J., Rao, R. P. N., Hayhoe, M. M., & Ballard, D. H. (1997). Eye movements reveal the spatiotemporal dynamics of visual search. *Psychological Science*, *8*(6), 448–453.

Zheng, L. (2016). The effectiveness of self-regulated learning scaffolds on academic performance in computer-based learning environments: A meta-analysis. *Asia Pacific Education Review*, *17*(2), 187–202.

Zheng, L. J., Mountstephens, J., & Wi, J. T. T. (2020). Multiclass emotion classification using pupil size in VR: Tuning support vector machines to improve performance. *Journal of Physics: Conference Series*, *1529*(5). 1–7.

Zima, E., Weiß, C., & Brône, G. (2019). Gaze and overlap resolution in triadic interactions. *Journal of Pragmatics*, *140*, 49–69.

Ziv, G. (2016). Gaze behavior and visual attention: A review of eye tracking studies in aviation. *The International Journal of Aviation Psychology*, *26*(3–4), 75–104.

Zwaan, R. A., & Radvansky, G. A. (1998). Situation models in language comprehension and memory. *Psychological Bulletin*, *123*(2), 162.

GLOSSARY

Afterimage A pattern of activation of the photoreceptors in the eye after a stimulus has been fixated, which persists after the stimulus has been removed. Particularly common with bright stimuli.

Anisocoria A condition where the pupils in each eye are of a different size.

Aqueous humor The aqueous humor is located in the front part of the eye, between the cornea and the iris. It consists mostly of water and is thus transparent.

AOI Areas of interest (aka ROI: region of interest): User-defined regions of the visual field that are of particular relevance to the researcher.

Aversion Gaze aversion is the movement of the eyes away from the face of the other speaker(s). The avoidance of eye contact.

Axon The elongated part of the nerve cell connected to the nucleus. The nerve signal travels along the axon.

Ballistic A fast movement of the eyes that cannot be controlled after it has been initiated. Saccades are ballistic.

Blind spot An area of the retina in which the ganglion cells' axons are gathered and leave the eye as the optic nerve. There are no cones or rods in this area, hence no vision takes place in the area.

Chemical transmission The transmission of information across the synaptic cleft. The firing of a presynaptic cell triggers the production of chemicals, which cross the synaptic cleft and reach the postsynaptic cell.

Ciliary muscles The ciliary muscles are two muscles in the eye that can alter the shape of the lens and thus help it to focus.

Circadian rhythm The circadian rhythm is the alternation of sleep and wakefulness in the body. The term originates from the Latin *circa diem* ("around a day"), as the circadian rhythm roughly matches a day.

Cohort aka word-initial cohort; all the words that begin with a given sequence of phonemes.

Cohort competitor Two words with the same initial phonemes: for example, "phone" and "fold."

Collagen A protein commonly found in the human body. Collagen is a component of the corneas.

Common ground Information that is available to all the participants in a conversation.

Compensatory movements Movements of the eyes to compensate for the movements of the head or the body and keep foveal vision stable.

Cones A type of photoreceptor cell in the retina. Cones work best in bright light and are less sensitive to light than rods.
Conjunctiva A transparent mucous membrane that covers the eye ball and the inside of the eye lids, but not the cornea.
Contrastive stress A prosodic marker indicating that the stressed item is contrasted against another item.
Convergence The process whereby cells receive input from many other cells.
Converging lens A convex lens.
Cornea The transparent part of the eye that covers the pupil in the front of the eye.
Corpus callosum The bundle of nerves centrally located in the brain, which joins the two halves of the brain.
Covert attention Attention directed to a location different from the foveal fixation.
Dead time The point of no return after a saccade has been planned when the saccade cannot be stopped but can be slightly altered.
Decoupling hypothesis The severing of the link between attention and foveal fixation. The link is decoupled during mind wandering, for example.
Decussation The crossing over of the optic nerve.
Deictic A pointer; pointing. In linguistics, deictics are words such as "here" or "this," which require a reference to the extralinguistic context. The deictic function of gaze is seen when one speaker follow's another speaker's gaze.
Dendrites Extensions of the nerve cell nucleus that allow the reception of synaptic messages from other cells.
Disfluency An error during speech: hesitations, false starts, filled pauses, word repetitions, slips of the tongue, stutters, and so on are all disfluencies in speech.
Divergence In neurology, the process whereby a cell sends its output to many other cells. Divergence is the opposite of convergence.
Drift A type of miniature eye movement. The tendency of the fovea to slowly move away from the locus of the fixation. It is random and very slow.
Dwell time Dwell time is defined as the amount of time that a user spends in a given location or within an AOI/ROI.
Dyadic An interaction in which there are two active participants.
EEG Electroencephalography is the measurement of the electrical activity of the brain, usually done with electrodes attached to the scalp.
EFL English as a foreign language.
Electrodermal activity Electric current or potential on the skin.
Electrooculography The measurement of the voltage around the eye, with electrodes placed on the skin.
ELF English as a lingua franca.
Epithelium The outmost layer of the cornea, which consists of a very thin layer of cells.
ERP An event-related potential is a measurable change in electrical activity pattern in the brain as a response to a stimulus (see: EEG).
ESL English as a second language.
Extraocular muscle The muscles governing the movements of the eye. They are called extraocular because they are outside the eye itself.
Eye contact illusion The overestimation of the actual amount of eye contact between speakers.
Eye-mind theory The assumption that whatever the speaker is fixating on is the focus of their attention and vice versa, that the speakers will fixate on whatever visually represents their topic of attention.

Eye-mouth continuum An individual in a face-to-face interaction distributes their fixations more or less evenly between the eyes and the mouth.

Eye movements The eye can move in three major directions: up and down, left and right, and torsionally. The movements are controlled by the extraocular muscles

FACS The facial action coding system; a system of notation for facial expressions.

FEF The frontal eye field is an area of the brain located in the frontal cortex. It controls the moving of the eyes, among others.

First fixation duration It is a subset of fixation time, which applies only to first fixations; it measures only the duration of the first fixation in an AOI.

Fixation density This metric correlates the number of fixations in a given AOI with its size. Closely related metrics are fixation coverage and fixation dispersion.

Fixation time Fixation time is dwell time, minus the time spent in saccades. Also known as fixation duration.

Fixations Fixations are relatively static periods during which the fovea is fixed onto a small area.

fMRI Functional magnetic resonance imaging.

Fovea The fovea (fovea centralis) is a depression in the retina in the macular region. It has the highest acuity of vison and is the area where the image is focused, when the eye is functioning normally.

Fundus The inner lining of the back of the eye.

Ganglion cell Photosensitive cells in the retina. Their axons form the optic nerve.

Gaze cross-recurrence A measure of how often two participants look at the same target.

Gaze paths See: scan path.

Glial cells Glial cells coat the axons of nerve cells. Largely made up of myelin, glial cells support, feed, and protect the cells from pathogens.

Glint A reflection of light in the eye.

GSR Galvanic skin response: a type of electrodermal activity. Galvanic skill response measures very small electric current on the skin, usually in the palm of the hand. It is very closely related to the sweat glands activity and hence it provides a direct way to assess one's emotional arousal and cognitive load.

Hz Hertz. The unit of measurement of frequency. One hertz is defined as one cycle per second. In eye tracking it is used to measure how often the eye tracker records the position of the eyes.

IC Inferior colliculus: the lower area of the colliculus (see: superior colliculus). It is responsible for hearing.

Implicature In Gricean pragmatics, a meaning conveyed by the speaker that is not stated explicitly in the sentence uttered by the speaker.

Inattentional blindness A perceptional phenomenon whereby under heavy cognitive load, stimuli that would otherwise be noticed are ignored. It is best known as the "invisible gorilla" experiment.

Intraparietal sulcus The intraparietal sulcus is located in the parietal lobe in the dorsal (superior) posterior (back) part of the brain. It is involved in visual attention, saccades, perception of depth in stereopsis, and hand-to-eye coordination.

Iris A thin membrane separating the front and back of the eye. The pupil is the central opening of the iris. The iris is the colored part of the eye.

IVT see Velocity-threshold identification.

Joint attention A form of social attention, in which both subjects are gazing at the same item and *one* of them is aware of both gazes. Opposed to shared attention, in which both subjects are mutually aware of the gaze.

Koniocellular Nerve cells that are smaller than parvocellular cells.

L1 The first language, or native language, as opposed to other non-native languages (Lx) acquired later in life.

Labile stage The period when a saccade has been started but has not been executed yet, so it is still possible to change it or cancel it altogether. The opposite of the labile stage is the nonlabile stage, aka, saccadic latency.

Lacrimal glands The lacrimal glands secrete the tear fluid.

Lasik surgery Laser-assisted in situ keratomileusis: that is, a type of surgery in which the cornea is reshaped (keratomileusis) using a laser. It is commonly used to correct myopia and other vision problems.

Latency (1) The time elapsed between the perception of light and the contraction of the pupil. (2) The period after a saccade has been initiated in which the eye is unable to perform another saccade.

Lateralization The specialization of one hemisphere in the brain over the other to perform some tasks. For example, the left hemisphere is associated with syntax, whereas the right hemisphere is associated with prosody.

Lens The transparent element of the eye that focuses the light rays onto the retina. It is between the aqueous humor and the vitreous body, behind the iris and the pupil.

LGN Lateral geniculate nucleus: A structure in the midbrain where the optic tract terminates.

LIP Lateral intraparietal area: an area within the intraparietal sulcus, which contributes to the control of saccades.

Macula A darker pigmented area of the retina about 5 mm in radius. It is responsible for sharp, clear, color vision. It contains the fovea, the parafovea, and the perifovea.

Magnocellular Larger cells in the LNG. The magnocellular cells receive information from rods. (see: parvocellular)

McGurk effect An interference between the visual and auditory channels (that results in the fusion of the two stimuli). For example, hearing the auditory stimulus for [p] while seeing the visual stimulus for [k] (i.e., the facial movements) results in the hearing of [t].

Mean fixation duration This represents the average length of fixations, usually in an AOI.

Melanin The brown or black pigment in skin and hair.

Membrane The external layer of a cell.

Microsaccades Also known as "flicks," they involve movements of the eye of less than 1 degree. They are, as the name indicates, usually smaller than saccades but otherwise very similar to them. See: miniature eye movements.

Mind probe Questions by the investigator to a participant in an experiment to check whether they are on task.

Mind wandering A disconnect between attention and gaze. Unlike distraction, no external stimulus causes the attention to drift off from the task at hand.

Miniature eye movements Eye movements which occur within a fixation. They are very small and generally remain within the foveal area. See: tremor, drift, microsaccade.

Modularity The theory of cognition that sees particular aspects of language and cognitive processing (e.g., syntax or lexical activation) as encapsulated modules that operate exclusively on a given input with no further input from the rest of the system.
Motor mimicry See synchrony.
Mutual gaze Participants in an interaction looking at each other's faces. Eye contact.
Myelin Myelin is primarily composed of fatty cells and protein. It is the principal component of the myelin sheath that surrounds the nerve cells' axons. It is what gives the brain and spinal cord's white matter its distinctive color.
Myopia Nearsightedness.
Name-related gaze The gazes produced by speakers when they are looking at an item they are about to name. Name-related gazes are longer than gazes for object recognition.
Neurotransmitters Chemicals used by the cells to transmit the neural signal across the synaptic cleft.
Oculomotor error Targeting errors that occur while executing a saccade.
Oculomotor nerve The nerve connected to the eye muscles that control the movements of the eyes, including the saccades.
Optic chiasm The optic chasm is the crossing over the nerves coming from each eye cross in a "X" shape (hence, the name, from the Greek letter X) so that some, but not all, of the information from the left eye goes to the right side of the brain and vice versa (see: decussation).
Optic nerve The optic nerve carries the output of the retina to the brain.
Optic radiation The fanning out of the optic pathway from the lateral geniculate nucleus into the occipital lobe.
Optimal viewing position The position that affords the best opportunity to process a word. The optimal viewing position is the center of the word or slightly to the right of the center of the word.
Parafovea The area surrounding the fovea. The parafovea has less visual acuity than the fovea.
Parafoveal preview While maximum visual acuity occurs in the foveal region, some degree of information can be captured by the parafovea. In reading the parafoveal preview may allow a reader the option of skipping the next word, especially if it is a short function word.
Parvocellular Smaller cells in the LNG. The parvocellular cells receive information from cones (see: magnocellular).
Perceptual span The region in the visual field from which information may be extracted. It includes the visual span and the parafoveal vision as well. The perceptual span extends fourteen to fifteen characters to the right and four to the left of the fixation.
Perifovea The area surrounding the parafovea. It is the periphery of the foveal region and has the least visual acuity.
PGN Perigeniculate nucleus. A part of the geniculate nucleus.
Photons A particle without mass that is a quantum of light. It travels at the speed of light, in a vacuum.
Photoreceptors The cells in the retina that are specialized in converting light into biological signals, and specifically electrical signals, at the membrane of the cell. Photoreceptors are also known as membrane voltage or membrane potential.

Photosensitive cells: Cells that are capable of perceiving light—the rods, the cones, and the ganglion cells.

Polyadic An interaction comprising more than two participants.

Population In statistics, a well-defined group of entities (objects, people, or behaviors) the researcher wants to be able to make a claim about, using a sample of the population. A population can be all the students enrolled in a high school or all US residents; all the cans of peaches canned by a food producer or all vehicles produced in the United States in 2010; the scores of all the test takers of the TOEFL in 2019 or the essays produced by students in one class.

Posterior occipital lobe The back part of the back of the brain.

Postsynaptic In chemical transmission, the cell that is receiving the signal.

Presbyopia Farsightedness.

Presynaptic In chemical transmission, the cell from which the signal comes from.

Prosopagnosia The incapacity to recognize faces.

Pupil The pupil is the opening of the iris that determines the amount of light entering the back of the eye.

Pupillary hippus A spasmodic contraction and dilation of the pupil not related to illumination. It is usually not pathological.

Pupillary light reflex The process of the pupil's contraction and dilation caused by the amount of ambient light.

Pupillometry The science of the measurement of the size of the pupil. Pupil dilation, aside from the light reflex, has been shown to be sensitive to cognitive effort and cognitive arousal.

Purkinje image Reflections of light in the eye, named after Jan Evangelista Purkinje (1787–1869). There are four reflections. See: glint.

Pursuit movements The movements of the eye, which occur when the eye follows a moving target. These movements include anticipatory or predictive components.

RAGE Reduced ability to generate expectations: the hypothesis that Lx learners cannot make as many predictions as L1 speakers because of reduced competence.

Refixation See: fixation. It takes place when an element is fixated sequentially more than once.

Regression See: fixation. A subject fixates on point *a*, moves on to point *b*, and then returns toward or to point *a*. Regressions denote processing problems or just repairs for overshooting the target in a saccade.

Retina The part of the eye where the focused imaged produced by the lens is projected.

Return sweep A return sweep is a long saccade executed when a reader has reached the end of the line in reading. The return saccade takes the reader to the beginning of the line below.

Rods Rods are a type of photoreceptor cell in the retina. They are elongated and work best in dim light, as they are extremely sensitive to light. Rods are most sensitive to movement and shapes and cannot detect color.

ROI Region of interest. See: AOI.

Saccade A fast, ballistic movement of the eyes. During saccades viewers are effectively blind. The purpose of saccades is to redirect the fovea to a new region.

Saccade rate A metric of the frequency of saccades.

Saccadic latency See: latency.

Saccadic suppression During a saccade, the movement of the eyes is too fast for the retina to be able to process the images. Not to be confused with saccadic dead time.

SC See: superior colliculus.

Scan path A representation of the visual scanning of a scene, consisting of a series of fixations, separated by numbered saccades. Generally, the fixations are represented by circles proportional to the length of the fixation and saccades are represented by straight lines connecting the circles.

Sclera The white, opaque external part of the eye that is visible (the white of the eye).

SEF Supplementary eye field: an area of the brain involved in the control of saccades.

Shared attention A form of attention in social interactions, in which both subjects are mutually aware of the gaze. Opposed to joint attention, in which only one participant is aware of both gazes.

Social gaze The communicative uses of gaze and attention.

Spotlight theory A theory of visual attention, which claims that attention is directed to a subset of the visual field. See: zoom lens theory.

Stereopsis Binocular vison.

Stereoscopic vision Vision with two eyes. Stereopsis is responsible for sense of depth (how far away objects are).

Striate cortex A part of the brain. The striate cortex is also known as the visual cortex.

Stroop effect A cognitive dissonance that arises from the fact that participants are asked to read color words, which may be printed in a congruent color (e.g., the word blue, in blue letters) or incongruent color (the word blue, in red letters).

Superior colliculus An area of the midbrain largely responsible for the targeting of objects of interest by saccades.

Symmetrical vergence See: vergence.

Synaptic cleft The gap between the axons and the dendrites.

Synchrony The temporal alignment of behaviors by the participants of a social encounter. Synchrony ranges from prosodic and grammatical features to gestures and stance.

TESOL Teaching English to Speakers of Other Languages.

Theory of mind The attribution of mental states to other people.

Topographical organization Organized on the basis of spatial location. The cells that are located near to one another are connected and these connections based on the topography of the cells are preserved along the visual pathway of the brain.

Torsional movements A type of eye movement that can be either clockwise or counterclockwise.

Tremors A type of miniature eye movement. They consist of very fast oscillations superimposed on drifts. They can be visualized as a zig-zag motion that occurs during the drift.

Triadic An interaction involving three participants. Depending on the definition being used, the third participant may be an object.

TRP Transition-relevant point is a technical term from conversation analysis, indicating a point in the conversation where a change in speaker may take place.

Velocity-threshold identification (IVT) An algorithm to identify saccades.

Vergence Asymmetrical movement of the eyes, usually to accommodate objects closer to the eyes.

Visual acuity The capacity to differentiate visual objects (e.g., letters in reading).

It is highest in the foveal region.

Visual cortex aka striate cortex. Part of the visual system. The visual cortex is the primary center of the brain for visual processing. It is located in the back of the brain.

Visual span The number of characters that are identifiable within one fixation, in reading.

Visual system The parts of the body that make vision possible, aside from the eyes; the human visual system is part of the nervous system (retina, optic nerve, optic tract, and the visual cortex).

Vitreous humor The vitreous humor is a transparent gel, composed of 99 percent water, which fills the eyeball.

VWP Visual world paradigm.

Zoom lens theory This theory claims that the focus of attention is variable, precisely like a zoom lens.

INDEX